Australia's **Lost World**

A History of Australia's Backboned Animals

Tom Rich and Patricia Vickers-Rich are vertebrate palaeontologists—they study the history of animals with backbones. Leaellyn Rich is their daughter, who has been raised in a world of books, fossil *Diprotodons* on the kitchen table and expeditions to remote parts of Australia and elsewhere in the world since she was 3 months old. She is now an Arts/Law student at Melbourne University. Tom is Curator of Vertebrate Fossils at the Museum of Victoria in Melbourne and Pat is Professor of Palaeontology at Monash University, also in Melbourne.

AF616700

Australia's **Lost World**

A History of Australia's Backboned Animals

Patricia Vickers-Rich
Professor in Earth Sciences and Ecology/Evolutionary Biology
Monash University, Melbourne

Leaellyn Suzanne Rich
Previous student MacRobertson's Girls High School;
Current Arts/Law Melbourne University

Thomas Hewitt Rich
Curator of Vertebrate Fossils
The Museum of Victoria, Melbourne

Kangaroo Press
in association with
Monash Science Centre
Monash University

Published with the assistance of the
Monash University Publications Committee

© Monash Science Centre 1996

First published in 1996 by Kangaroo Press Pty Ltd
3 Whitehall Road Kenthurst NSW 2156 Australia
P.O. Box 6125 Dural Delivery Centre NSW 2158
Designed and Typeset by
Pioneer Design Studio Pty Ltd Lilydale 3140
Printed in Hong Kong through Colorcraft Ltd

ISBN 0 86417 798 4

Cover illustration by Peter Trusler,
courtesy of *Australia Post*

Dedication

To Kevin Broadrib, who had faith in this book for many years before it materialised.

PREFACE

A curious five-year-old can be an awe-inspiring inquisitor. A few years' headstart on that five-year-old allows a parent to answer or at least give some hints to the questioner. A common defensive action is to refer to source books for further information. But a few years ago when our daughter began asking questions about Australia's prehistoric animals, some of which we were working on, the show-and-tell session ended with that conversation or a quick trip to the museum collection where one of us worked. Few, if any, books existed on the topic of Australia's prehistoric backboned animals, and if they existed at all, they were not written for children. The result of this challenging period of parenthood and childhood is *Australia's Lost World,* written by all of us, parents and daughter together. It is a book aimed mainly at upper primary and early secondary students, because that is the stage to which most of our family has progressed to date. We hope, however, that adults and younger children (our six-year-old Timmy included!) will be able to enjoy many parts of the book as well, be it through the pictures alone or in its entirety. We encourage parents who might be reading the book to younger children to use parental licence and embellish your tale-telling with stories made up from the facts presented!

CONTENTS

PREFACE **5**

ACKNOWLEDGMENTS **8**

CHAPTER 1: IMAGINING **9**
An Introduction

CHAPTER 2: THE WAY THE WORLD USED TO BE **14**
The Last 4000 Million Years of the Earth's History

CHAPTER 3: ONE-CELL BLOBS **31**
Single Cells, the Origin of Life, the First Animals and Plants and Their Multitude of Cells

CHAPTER 4: ICE AND EXPLOSIONS **38**
Ice Ages, Many More Cells, An Explosion of Armoured Life

CHAPTER 5: SIFTERS IN A TROPICAL PARADISE **46**
Australia's First Animals with Any Backbone

CHAPTER 6: REVOLUTION OF THE JAWS **57**
Fish, Fish and More Fish.
Attack on the Tropical Seas and Streams

CHAPTER 7: LAND AHOY! **67**
Plants and Amphibians "Slither" onto Land

CHAPTER 8: REFRIGERATION DE LUXE, AN ANTARCTIC ADDRESS **72**
Australia Cruises to the South Pole. The Late, Great Palaeozoic Ice Age, and Then Someone Turns on the Heater

CHAPTER 9: ENTER THE DINOSAURS **81**
Australia's Successful Failures: Polar Dinosaurs, Cool Sea Monsters, and Big Troubles 65 Million Years Ago

CHAPTER 10: RAINFORESTS TURN TO DUST **100**
Australia Parts Company with Antarctica: Cool, Wet Forests, and Flamingo Covered Lakes of the Centre. The Great Drying Out, Our Deserts Begin

CHAPTER 11: TOMORROW **124**
Australia's Face and Place in the Future.
Kangaroos and Pandas, Eucalypts and Birch - Which Will Win?

SOME REFERENCES FOR FURTHER READING **128**

ACKNOWLEDGMENTS

We would like to thank several people for the help given which made this project possible. Kevin Broadrib, to whom this book is dedicated, had unfailing faith in the book's completion over a long period of time. We thank him for being a very patient editor. We would also like to thank Ray Cas, who orginally introduced us to Kevin, an introduction which has led to the production of several children's books. Dawn Titmus and Shelley Kenigsberg have also been great help in commenting on manuscript, as has Carol Natsis. We thank the Smithsonian Institution and John Gurche for allowing us to use illustrations from *The Tower of Time* mural. Draga Gelt was of invaluable assistance in preparation of the graphics used throughout the book and Steve Morton and Frank Coffa provided much of the photographic material. A number of others provided individual photographs, and they are acknowledged in the figure captions. Our sincere thanks to each of these people. Especial thanks are owing to Frank Knight, the Museum of Victoria (Melbourne) and Pioneer Design Studio for use of illustrations originally published in *Kadimakara: Extinct Vertebrates of Australia* (P. V. Rich and G. F. van Tets, eds., 1985). Monash University and the Museum of Victoria, and especially Professors Ray Cas, Jim Cull, Gordon Lister and James Warren, Dr. Robert Edwards and Dr. Jim Bowler, provided essential resource facilities where much of this book was written, and both institutions have given full support to making aspects of current research of their staff available to the public. For that generous attitude we are most grateful. Francis de Souza was most helpful in advising us on computer matters. Also to be thanked are the numerous funding agencies that have supported our research and field work over the last fifteen years, which have allowed the discovery and study of many of the animals discussed in this book: the National Geographic Society, the Australian Research Council, Monash University, the Museum of Victoria, the Friends of the Museum of Victoria, Atlas Copco, Mobil Oil, Imperial Chemical Industries, Safeway, David Holdings, the Ingram Trust, the Sunshine Foundation, the Danks Trust, Utah Foundation, the Ian Potter Foundation and the Victorian State Government (Employment Initiatives Scheme). And finally to our proof readers: Kerry Auslebrook, Jenny Monoghan, Kathy Smith, Gerry Kool, Mary Walters; and to our ever patient designer – Derrick Stone.

Patricia, Leaellyn and Tom Rich
Melbourne, November 1995

CHAPTER 1:

IMAGINING

An Introduction

A SPLASH IN A PALAEOZOIC SEA

Our home is in the middle of the Bush. We are surrounded by a Messmate Stringybark and Peppermint forest with undergrowth of *Banksias*, *Hakeas* and secretive, minute orchids which occasionally show their flowers above the soil. Sometimes an Echidna wanders through the yard, and the trees are often crawling with Crimson Rosellas or big, beautiful Black Cockatoos just before rain. Very rarely, a Wedge-tailed Eagle drifts overhead.

Figure 1 .1: The Bush around our house (a and b), has plants typical of modern Australia – *Banksias* (c), *Hakeas* (d), Bracken Ferns forming the undergrowth and a variety of eucalypt and wattle trees. These plants are very different from those known in other parts of the world, because Australia has been isolated from the other continents for more than 50 million years. The eucalypts (*Eucalyptus*) started out in Australia - they are real natives. The wattles and their relatives, *Acacia*, however, are rather new arrivals; they came from Asia, evidently a few million years ago, and have done very well indeed.

(a)

(b)

(c)

(d)

How unique to Australia are all these living things. Australians take them for granted, but they are animals and plants of wonder to a girl growing up in California or a boy grabbing at tadpoles along the Amazon River in Brazil. Australia's kangaroos and koalas have long been symbols of this country, be they in a first year reader or on the tail of a big Qantas 747 heading down the runway in Hong Kong.

Figure 1.2: The present-day Emu (*Dromaius novaehollandiae*) is at the end of a long line of birds that are unique to Australia. Fossils from birds dating as far back as 15 million years are closely related to the Emu. But which birds outside of Australia and New Guinea are nearest cousins to the emu and its close relatives, the cassowaries, is still an unanswered question - although there are many theories! The emu is one of the many animals and plants unique to Australia which show that the continent has been isolated from the rest of the world for a long time.

Figure 1.3: Our unusual animals have become symbols for Australia, as this QANTAS 747 travel documents folder shows. Our isolation from the rest of the world, even now, is made clear to us by the fact that we must board a plane (or a boat) to get out of Australia. Our country is surrounded by water; it is still an island continent, but that situation may not last forever.

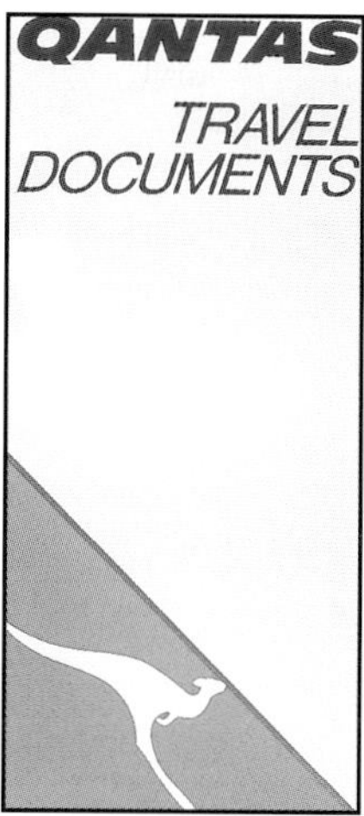

Figure 1.4: The city of Hong Kong, gateway to Asia, is about a seven hour flight from Sydney, and its animals and plants and those of neighbouring mainland China are worlds apart from those of Australia. Animals and plants in Indonesia, one of our nearest neighbours, are also very different from Australia's.

Figure 1.5: The Great Wall of China crosses a land not all that far from Australia but which has very different backboned animals. However, at times in the past, such as in the Devonian Period, some 400 million years ago, vertebrates of these two continents were very similar. Then the two continents were close enough for animals to move back and forth between them - something that may well happen again in the future, about 40-50 million years hence.

Why are our plants and animals so very different from those in the rest of the world? This seems to have been caused by Australia's long isolation from other continents. It is a few hours flight to China and Japan, and more like 18 hours to North America, even longer to Europe. Water surrounds and separates Australia from all these other places now, as it has done for several millions of years. There have been a few times in the more recent past (approximately 10,000 years ago), however, when places like New Guinea and Tasmania were actually connected to the Australian mainland.

Figure 1.6: The Giant Panda, a native of Asia, may some day meet the kangaroos – that is if both groups survive the insults of humans and changing climate. Australia is moving north at a rate of about 10 centimetres per year. In about 40 to 50 million years from now, if that northward movement continues at the same rate, Asia and Australia should dock alongside each other. Who will win when the bamboo-chomping pandas (if they survive) and the browsing and grazing kangaroos begin competing for the same food and living space? The selection that nature makes will determine which animal survives! (M. Chow)

Even weirder and more fantastic than the animals we have in Australia today are those many more that lived in the past and are now extinct. Just imagine a giant, trunked sloth-like marsupial trudging and grunting from tree to tree, ripping off great strips of bark with massive clawed hands; or a cheetah-sized carnivorous rat kangaroo in hot pursuit of an emu, which it eventually drags down for its supper; or a long, snorkel-snouted, armour-plated fish swimming around strange-looking corals on a reef that grows where the Kimberleys now poke up their heads in a dry desert. Just imagine....

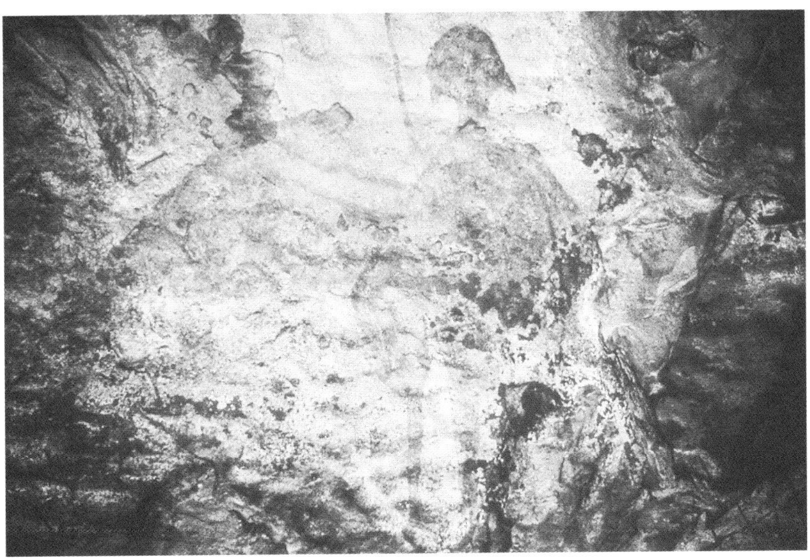

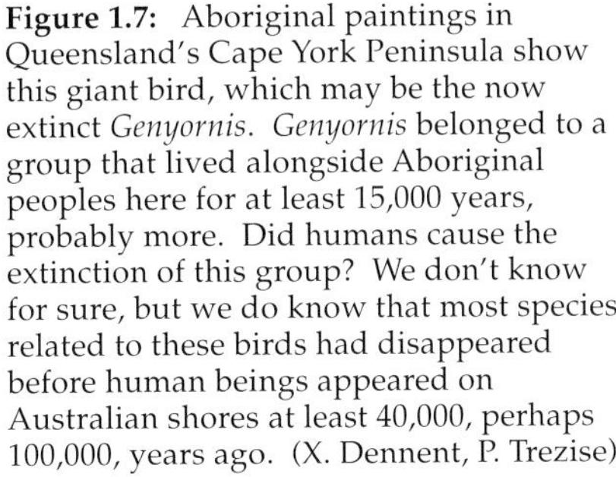

Figure 1.7: Aboriginal paintings in Queensland's Cape York Peninsula show this giant bird, which may be the now extinct *Genyornis*. *Genyornis* belonged to a group that lived alongside Aboriginal peoples here for at least 15,000 years, probably more. Did humans cause the extinction of this group? We don't know for sure, but we do know that most species related to these birds had disappeared before human beings appeared on Australian shores at least 40,000, perhaps 100,000, years ago. (X. Dennent, P. Trezise)

This book is full of *imagining* - but not just ordinary *imagining*. It has some realness. Our *imagining* of the past in Australia is based on real fossils that have been found over the years since at least the early 1800s. Our *imagining* is based not only on the fossils, but also on the study of those fossils by palaeontologists, who are trained to interpret them based on their knowledge of living animals, plants and environments. Our *imagining* comes from a solid base – our understanding of the modern world, which allows us to go back to the past as well as to make predictions about the future.

Figure 1.8: *Palorchestes* was a giant tree-ripping marsupial that lived in Australia during the last five million years. Now extinct, it was one of the weirdest marsupials this continent has produced. *Palorchestes* probably had a trunk, unlike any marsupial living today, and massive, powerful arms and hands with large claws. (From Rich, van Tets and Knight, 1985)

Figure 1.9: *Propleopus* was something more than your average kangaroo. Instead of grazing on grass or browsing on leaves, *Propleopus* was probably a carnivorous kangaroo, a fleet-footed killer that was Australia's answer to the cheetah. It may have been Australia's only carnivorous mammal able to pursue prey in open country, where speed is all important in getting a day's lunch.

Figure 1.10: Some of the places where fossils of backboned animals have been found in Australia. (Modified from a map in Rich, van Tets and Knight, 1985)

CHAPTER 2:

THE WAY THE WORLD USED TO BE

The Last 4000 Million Years of the Earth's History

As little as 10,000 years ago you could walk to Tasmania from Melbourne, or to New Guinea from the tip of the Cape York Peninsula. A few million years earlier you could have walked all over the Great Barrier Reef, and 280 million years ago you would have been able to stare at some gigantic glaciers in many different parts of Australia, such as at Bacchus Marsh, just west of Melbourne.

Australia has changed a lot over the last 4000 million years of its history. It has looked as it does today for only a few thousand years, and it has been an isolated island continent with its kangaroos and koalas for only say, 40 to 50 million years. Before that, it was part of a much bigger continent, a supercontinent called Gondwana, which also included Antarctica, South America and at times, Africa, India and even parts of southern China, so some people think.

How do we know all this is true? We owe these great stories to geologists - they are good "yarn-spinners" and even better detectives. Like detectives, they have to try to piece together the details of a geological "crime" from a few clues left behind, such as the composition of rocks - are they sandy, silty or lava-like? Geologists also look at what has happened to rocks over a period of time - have they been folded, or twisted, or are they lying flat and horizontal in relation to the present land surface? All these features can give geologists a hint of what has happened in a certain area. Then, by putting together all the bits and pieces of information from all over the continent and comparing it with what else is known from the rest of the world, they can put together an overall picture that explains what has gone on globally.

In this book, the last 500 to 600 million years of the Earth's history is examined. This is a really important time, because it has a good fossil record. From about 570 million years ago up to today, many animals and plants had some kind of a hard skeleton, and parts of these skeletons have remained behind as fossils. If an animal or plant doesn't have some sort of a protective skeleton, its chances of becoming a fossil are rather poor. We're going to look mainly at animals during this time, particularly those that had backbones, our immediate ancestors and their cousins, uncles and aunts. We won't completely ignore the rest of the life that existed alongside the vertebrates, but will concentrate on the backboned animals of Australia because these close relatives of ours are fascinating and should

Figure 2.1: A time scale of events over the last 4,600 million years. Procaryotes are life forms without a nucleus in their cells, like blue-green algae and bacteria. Eucaryotes are organisms whose cells have a nucleus. A nucleus is a place where genetic information is concentrated, where DNA (deoxyribonucleic acid) forms chromosomes. Redbeds are just that, red coloured beds of rocks. They can form only when oxygen is present in the water or the atmosphere. We need oxygen in order to breathe.

FA means first appearance, such as the FA of redbeds about 2600 million years ago. You might have been able to breathe then, but before that you would not have been able to live on the Earth without a space suit!

* indicates times of glaciation, when the Earth became very cold. The long glacial period between 800 and 600 million years ago was about the coldest time on our planet since it began. The Cretaceous Period was one of the warmest times, with a climate suitable for dinosaurs.

The line representing the sea level shows that during times of cold, during glacials, the sea level was low, because much sea water was frozen in ice at the poles, whereas during warm times sea levels were high. So, by looking at these sea-level curves you can guess how cold or warm the Earth was.

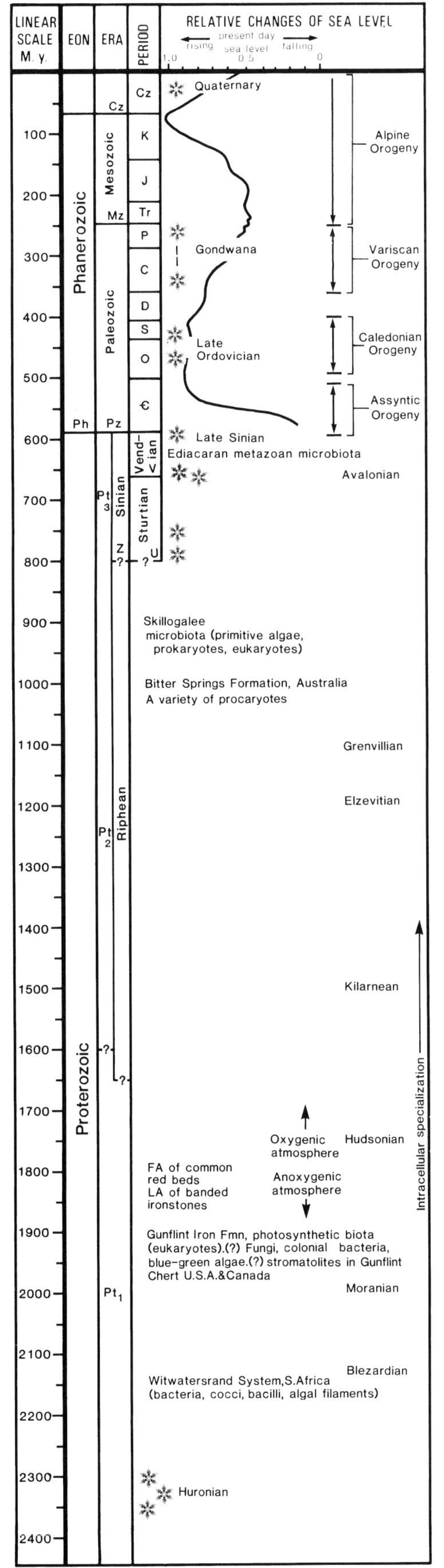
LINEAR SCALE M.y.
EON
ERA
PERIOD
RELATIVE CHANGES OF SEA LEVEL
present day sea level
rising
falling
Quaternary
Alpine Orogeny
Gondwana
Variscan Orogeny
Late Ordovician
Caledonian Orogeny
Assyntic Orogeny
Late Sinian
Ediacaran metazoan microbiota
Avalonian
Phanerozoic
Proterozoic
Mesozoic
Paleozoic
Sinian
Riphean
Vendian
Sturtian
Skillogalee microbiota (primitive algae, prokaryotes, eukaryotes)
Bitter Springs Formation, Australia
A variety of procaryotes
Grenvillian
Elzevitian
Kilarnean
Intracellular specialization
Oxygenic atmosphere
Hudsonian
FA of common red beds
LA of banded ironstones
Anoxygenic atmosphere
Gunflint Iron Fmn, photosynthetic biota (eukaryotes).(?) Fungi, colonial bacteria, blue-green algae.(?) stromatolites in Gunflint Chert U.S.A.&Canada
Moranian
Blezardian
Witwatersrand System,S.Africa (bacteria, cocci, bacilli, algal filaments)
Huronian

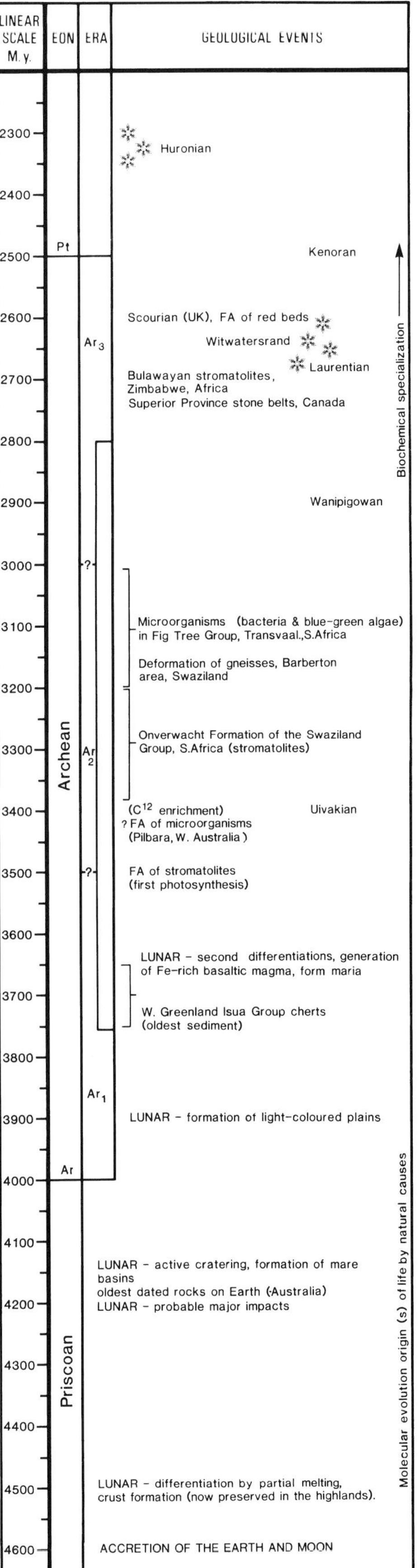
LINEAR SCALE M.y.
EON
ERA
GEOLOGICAL EVENTS
Huronian
Kenoran
Scourian (UK), FA of red beds
Witwatersrand
Laurentian
Bulawayan stromatolites, Zimbabwe, Africa
Superior Province stone belts, Canada
Biochemical specialization
Wanipigowan
Microorganisms (bacteria & blue-green algae) in Fig Tree Group, Transvaal.,S.Africa
Deformation of gneisses, Barberton area, Swaziland
Onverwacht Formation of the Swaziland Group, S.Africa (stromatolites)
Archean
(C12 enrichment)
Uivakian
? FA of microorganisms (Pilbara, W. Australia)
FA of stromatolites (first photosynthesis)
LUNAR - second differentiations, generation of Fe-rich basaltic magma, form maria
W. Greenland Isua Group cherts (oldest sediment)
LUNAR - formation of light-coloured plains
LUNAR - active cratering, formation of mare basins
oldest dated rocks on Earth (Australia)
LUNAR - probable major impacts
Priscoan
Molecular evolution origin (s) of life by natural causes
LUNAR - differentiation by partial melting, crust formation (now preserved in the highlands).
ACCRETION OF THE EARTH AND MOON

help us to see ourselves in a more sensible perspective. We are part of the animal kingdom, and are governed by its laws. We are by no means in complete control of things.

We'll take a walk (sometimes a swim!) back in time on (and near) the Australian continent, starting far back, say about 570 million years ago and amble across those millions of years right up to today. In the process, we hope that you will gain some idea, some feeling, for the changes that have happened on the Australian continent. In the last chapter we will go off into the future - using what we have learned from the past to predict just where we might be going, some 50 million years from today.

GEOLOGICAL LANGUAGE: A TIME SCALE

Just to gain your bearings, you will need a short course in geologist's language. By giving labels to certain periods of time (such as the Cambrian Period), geologists save a lot of time. They don't have to say "the time between 510-570 million years ago" whenever that particular span of years is talked about. They just say the Cambrian Period and know what it means. In the same way, the rest of the 510 million years of time is divided up into many different time spans - for example, the Ordovician Period from 510 to 439 million years ago, the Silurian Period from 439 to 408.5 million years ago, and so on. By looking at Figure 2.3 try working out when the Permian and Jurassic periods happened and then do the same for the other periods in Figure 2.3. If you can do this, and we are sure that you can, you are well on the way to becoming fluent in the GEOLOGICAL LANGUAGE.

For the early part of the Palaeozoic Era (which includes from oldest to youngest the Cambrian, Ordovician, Silurian, Devonian, Carboniferous, and Permian periods), much more of Australia was covered by ocean than it is now. The maps that are included in this chapter, called palaeogeographic maps (because they map out what Australian geography looked like in times past), show just where the oceans, reefs and land areas were once located in comparison to where they are now.

The Cambrian
590 to 510 million years ago

In the Cambrian Period more than half of the continent was covered with sea water of some kind. The whole south-east and much of the east coast of today was well and truly under water - deep water. Further west were shallower waters, like those in today's northern Queensland. In fact, there were even broad areas of very shallow and limey (full of calcium carbonate, just like on the Great Barrier Reef now) seas. Some just west of today's Alice Springs were so shallow that the water completely dried up and left salts on coastal salt pans, which have been preserved in the fossil record. Extensive reefs developed where the Flinders Ranges in South Australia are today. The

Figure 2.2: Different types of rocks.

(a) Sedimentary rocks
1. Sandstone **2.** Limestone
3. Conglomerate **4.** Siltstone
5. Diatomite **6.** Coal

(b) Metamorphic rocks
1. Green schist, formed in downgoing slabs of the Earth's crust in places near deep ocean trenches, such as under Japan today. **2.** Gneiss

(c) Igneous rocks
1. Basalt **2.** Granite

Sedimentary rocks are deposited by rivers, lakes, oceans and even by the wind. The size of particles in a rock gives an indication of how energetically or how fast the river or ocean current was flowing when the particles were deposited. Conglomerates require fast water; silts (which fossilise into siltstones) much slower water. The quiet waters of lakes and oceans away from the beach deposit mudstone and limestone. When rocks such as these are buried, heated up and bent by mountain building, *metamorphic rocks* such as schists and gneisses are formed. By studying them and the kinds of minerals they have in them, you can work out just how hot they became. Many of our precious mineral deposits, such as those at Mt Isa (Queensland) and Broken Hill (New South Wales) were formed in this fashion more than 1,000 million years ago. *Igneous rocks* are formed from molten rock material that cools after being thrown out of a volcano or while still underneath the ground. By studying the minerals and the size of the crystals in these rocks, you can work out many things about their history. For example, rocks such as basalts, with tiny, microscopic crystals, cooled very quickly; granites, with large crystals, cooled very slowly, thus allowing the crystals to grow large before the rock became solid. Igneous rocks are very helpful in allowing us to determine where continents were in times past. First, we can use the radioactive minerals, such as potassium and uranium, contained in the rocks to find out how old the rocks are. Then, we can measure the orientation of the magnetic minerals (those containing iron) in these rocks. As the rocks solidified, the iron minerals lined up according to where the north and south poles were at the time, acting like tiny magnets. By measuring this for Australian igneous rocks of the early Cretaceous age, for example, we can plot where the pole was relative to the position of Australia 110 million years ago. When we do this, we find that Melbourne was about 75-80 degrees of latitude south when dinosaurs were roaming around, rather than 38 degrees south, as it is today. So, the basalts can give us a clue to where Australia was millions of years ago.

(a) Sedimentary rocks **1.** Sandstone

2. Limestone

3. Conglomerate

4. Siltstone

5. Diatomite

6. Coal

(b) Metamorphic rocks **1.** Green schist

2. Gneiss

(c) Igneous rocks **1.** Basalt

2. Granite

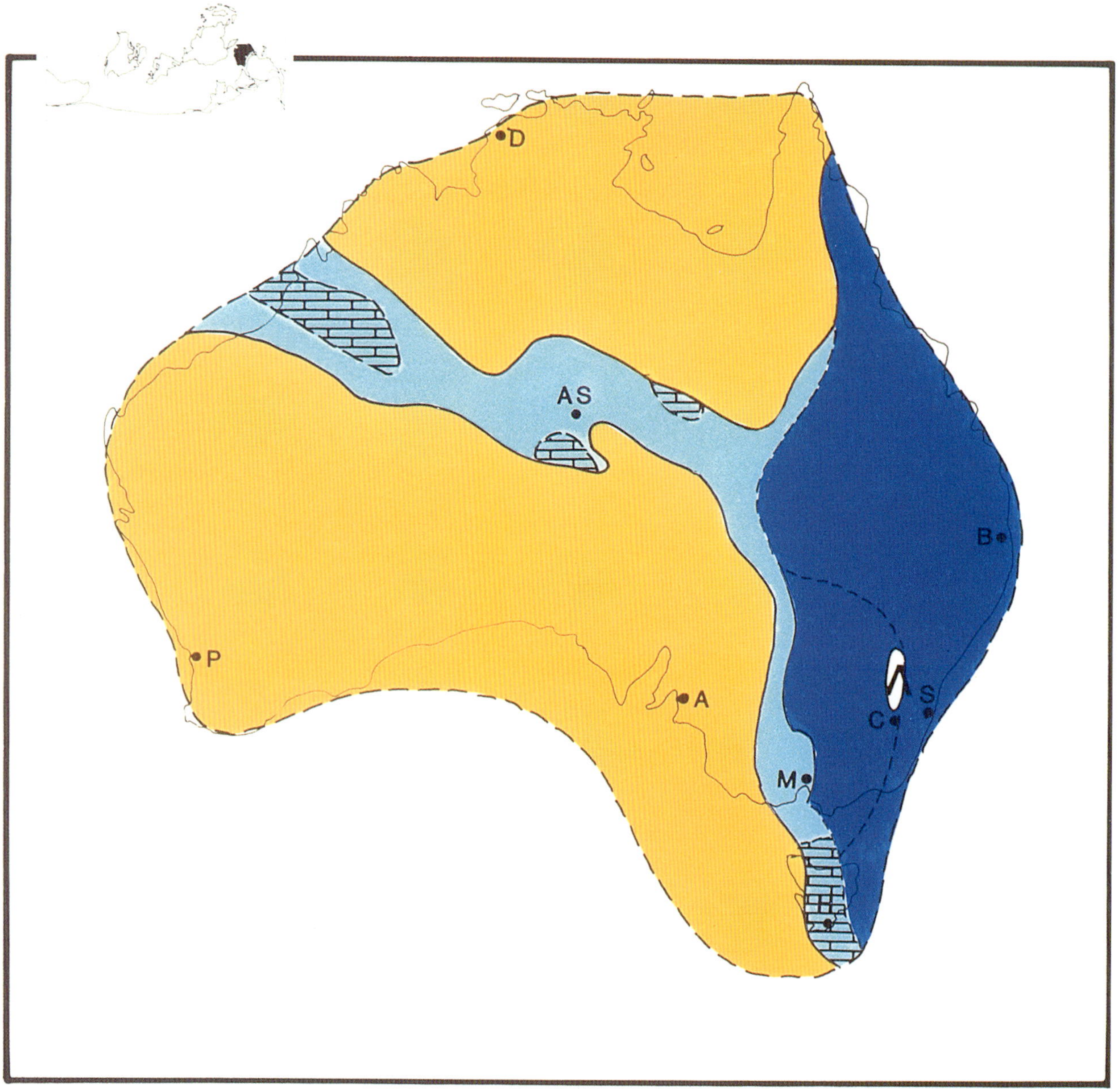

Figure 2.4: What the world, including Australia, was like 510 to 439 million years ago, during the Ordovician Period. In this map and all that follow these colours and symbols are used: Dark blue, deep ocean; light blue, shallow ocean; light blue with brick pattern, shallow carbonate-rich seas ideal for reef formation; green, swamps; gold, dry land areas; yellow, river deposits; wavy lines, reefs; black triangles, glacial deposits; bold upside down **V**'s, volcanoes; small v's, volcanic deposits.

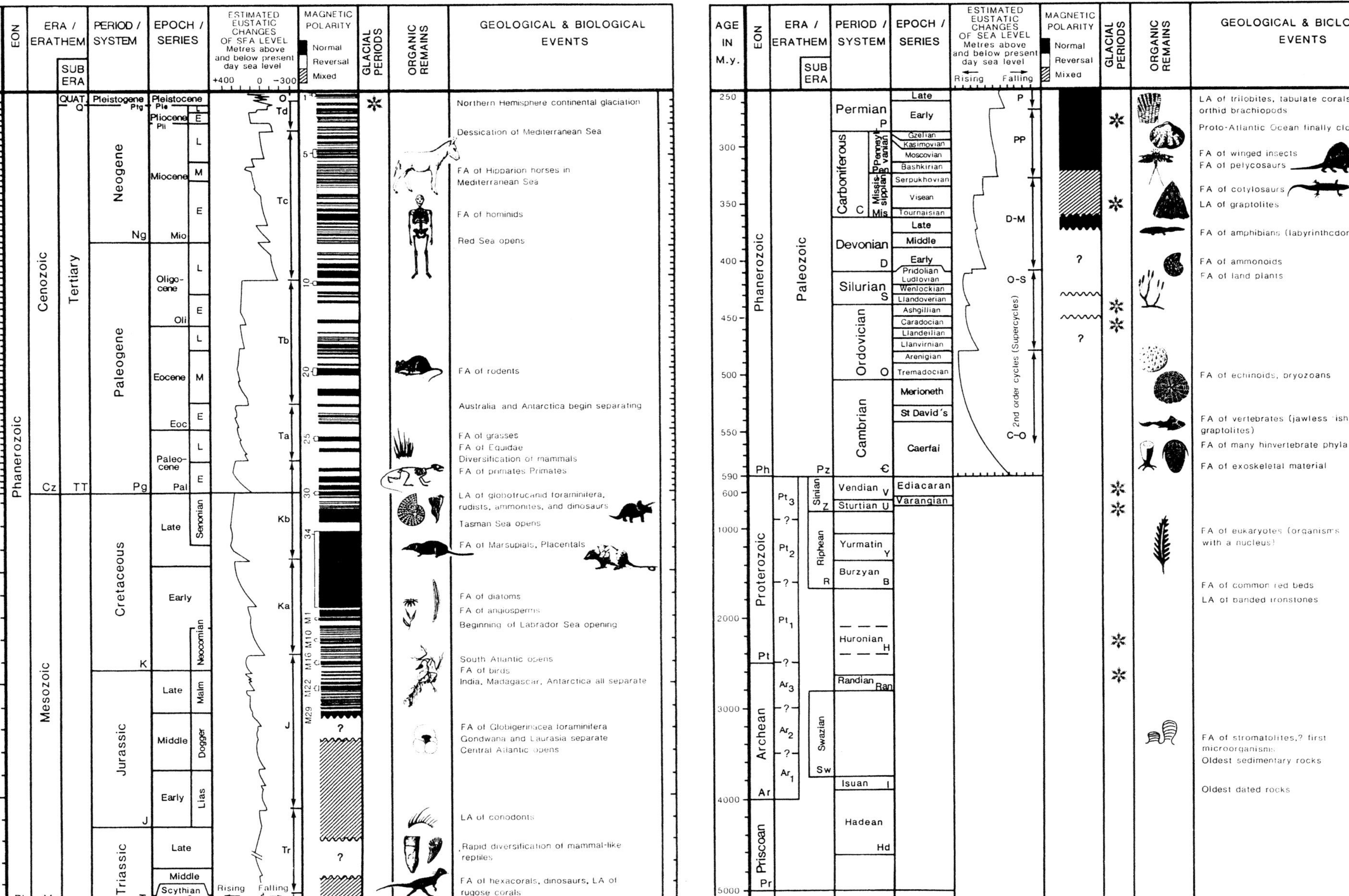

Figure 2.3: Detailed geologic time scale for the last 570 million years, a time-slice or eon called the Phanerozoic. This is the time during which animals and plants developed hard skeletal parts and so have left a reasonably good fossil record. Before this came Precambrian times (the Proterozoic, Archean, and Priscoan or Hadean eons), during which the fossil record is very poor, even though it is a much longer time period than the Phanerozoic.

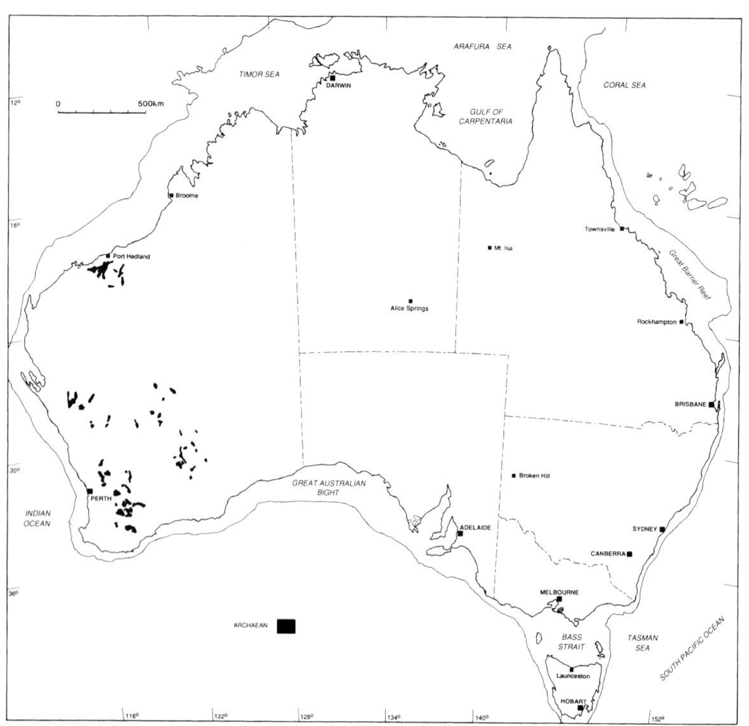

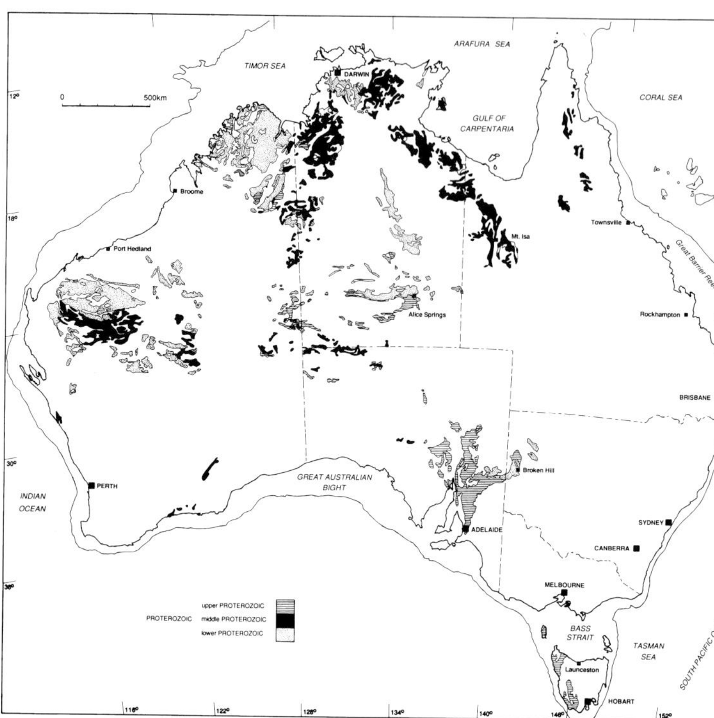

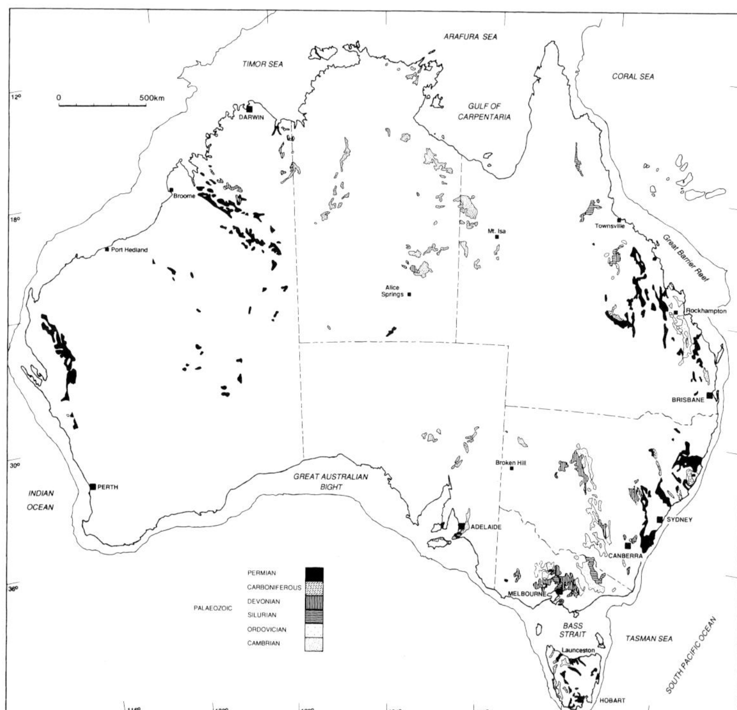

Figure 2.5: These five maps show where sedimentary rocks (those deposited as sediments in the sea, rivers, lakes and on land) and metamorphic rocks (those altered by heat and pressure) are known in Australia according to the geological period in which they were formed. Our oldest rocks from the Archean (4,000 to 2,500 million years ago) are found only in Western Australia, mainly in the northwest. These and those which formed during the Proterozoic (2,500 to 570 million years ago) contain many of our precious mineral deposits.

(a) Archean (4,000 – 2,500 million years ago).

(b) Proterozoic (2,000 – 590 million years ago).

(c) Palaeozoic (590 – 248 million years ago).

(d) Mesozoic. (248 – 65 million years ago)

(e) Cainozoic (65 – 0 million years ago).

seas were tropical, but not calm. Volcanoes were exploding all over the place. Huge piles of volcanic lavas and ash poured out to the north-west and south-west of Alice Springs. In Australia's south-east, there were chains of volcanic islands, just like those in Indonesia, Bougainville and the south-west Pacific today. Beautiful, but dangerous, to animals living in the area at that time.

Except at times of volcanic explosion, these dangerous

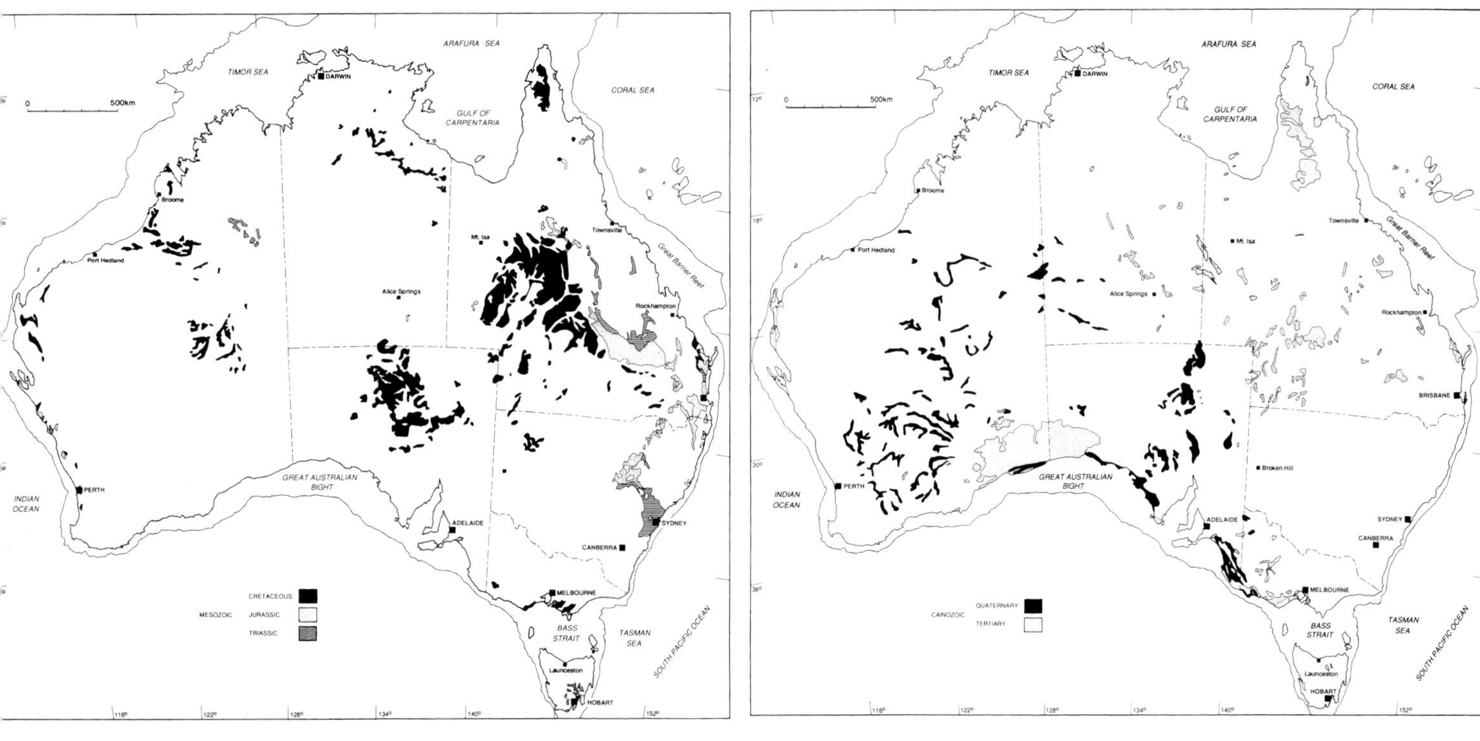

waters tended to be clear and clean, and many sandy rocks were laid down. Temperatures were warm; the whole of Australia at this time would have been like tropical Queensland of today. The trilobites, brachiopods and algae that were around at the time did very well, as we will see in Chapter 4.

The Ordovician
510 to 439 million years ago

All Australians **should** know about the Ordovician Period. It's the time when our earliest ancestors appeared on the scene. These were rather unobtrusive little forms - fish, one called *Arandaspis*, which we will get to know much better in Chapter 5. *Arandaspis* lived by swimming around slowly in the shallow, tropical seas of the time and sifting out micro-organisms, which were varied and abundant in the tropical climes - not a bad way of life at all. The palaeogeographic map of the Ordovician (Figure 2.5) shows that things had changed a bit from Cambrian times. The tropical seas were still present, but they didn't cover so much of our Australian continent. The deep oceans were confined to the east coast, and a rather narrow seaway split the continent into two, a northern part and a southern part. Volcanoes were not making life in the oceans as precarious as they had in the Cambrian, with only one major centre of activity situated somewhere down in the south-east - just north of where Canberra is today, and slightly to the west of Sydney.

The clear, shallow seas of this time were teeming with many kinds of animals that lived by filtering micro-organisms out of the silt-free waters, much more so than in the Cambrian, when most animals were mudgrubbers or bottom feeders of some

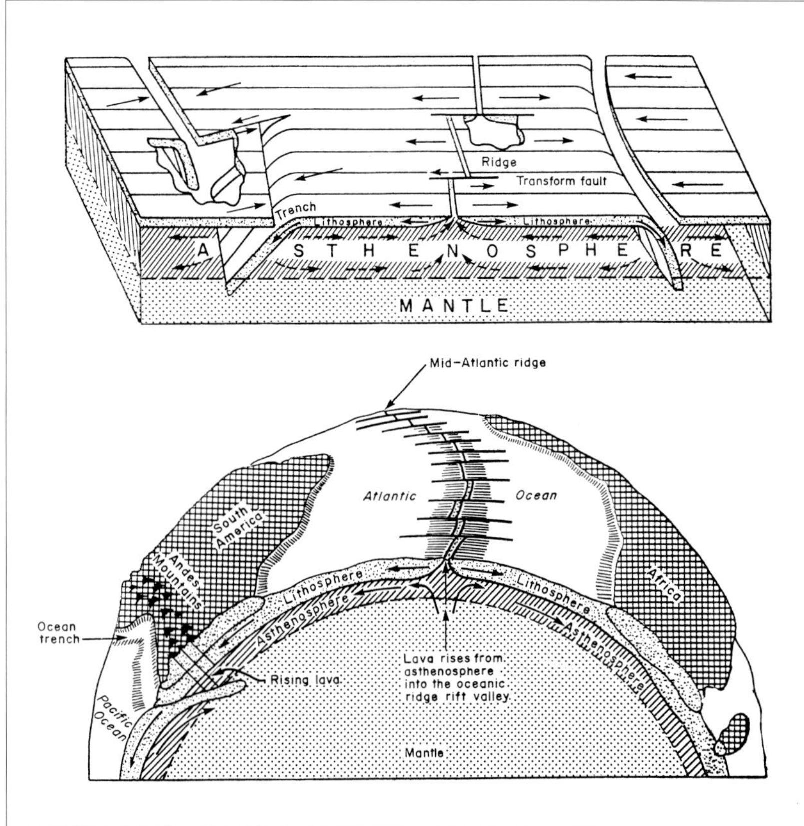

Figure 2.6: The way we think the Earth works is very different today from what we thought 30 years ago. It seems to be ever-changing; continents are not always in the same place, and over the last 500 million years, while vertebrate animals have been around, continents have moved quite a lot.

Plate tectonics is a theory that tries to explain how and why continents move. This theory says that the Earth's crust, the upper 70 kilometres or so, is divided into several big, rigid plates, which include both continents and ocean basins. The continents move because new hot rock is added along ridges, and this makes the plates on either side of ridges grow, for example along the mid-Atlantic Ridge or the ridge between Australia and Antarctica. Shallow earthquakes and volcanic activity occur along these ridges. Because the Earth doesn't seem to be expanding, if something is added in one place, it must be taken away somewhere else. As a result the crust dives back into the deeper parts of the Earth along trenches (also called subduction zones), such as in Japan or in the Kermedec Trench. Near such trenches big, shallow-to-deep earthquakes and lots of volcanic activity occur, such as the volcanoes in Japan or Indonesia. All this activity has been going on for some time, and it is what seems to explain continental movement. It allows constant recycling of crustal rocks and may be very important in keeping the right amount of carbon dioxide in our atmosphere. If it weren't for this activity, we might have the same kind of atmosphere as Mars, which would lead to an extremely cold temperature - bad news for us and most other life forms. Plate tectonics, and continental drift, and geologic recycling may be terribly important in making the Earth a place that has nurtured life.
(Figure courtesy of F. A. Middlemiss, P. R. Rawson, G. Newall, P. Wyllie and L. Sykes.)

sort. Just like the animals of the Cambrian, however, life was all in the water. The land was essentially lifeless, unless a few algae were just surviving in the sheltered 'soils'. If you had swum in the seas, you would have been amazed at the number of different kinds of life. A walk on land would have been monotonous for a palaeobiologist (one who studies ancient life). They would have been able to see all the rocks clearly - not one plant of any kind would have hidden them.

Figure 2.7 Some 500 million years ago in central Australia, slightly west of Alice Springs, shallow Ordovician seas may have looked like this. This picture was taken on the Great Barrier Reef a few years ago, but imagine yourself in shallow water like these waters, 500 million years ago.

The Silurian and Devonian
439 to 362.5 million years ago

Australia was still a tropical paradise during this period. It was still firmly attached to Antarctica and India, and these continents, in turn, were joined to Africa. The ocean gap between northern Australia and China was still not very wide, but it was growing. Something else was beginning to change too - less and less of the Australian continent was covered by shallow, or for that matter deep, oceans. In fact, if you look at the map for the Devonian (Figure 2.8), you will see that oceans were only just lapping on to the east coast and a little bit on the western and north-western part of the continent. In the Devonian, vast river systems flowed over parts of south-eastern Australia, such as around Mansfield and Buchan in Victoria, and in central and western New South Wales, and further north in southern and western Queensland.

A variety of heavily armoured, jawed fishes left their bones and armour plates in the sands and silts deposited by the streams and lakes - you will learn more about these in Chapter 6. Colourful reefs of sea lilies, corals, and brachiopods formed barriers to the open ocean around Buchan in Victoria and up in the north-western part of the continent. Snorkelling in the Devonian would have been just as good in those places as on the Great Barrier Reef today. The Great Barrier Reef we know did not exist then. There was just deep ocean where corals now grow.

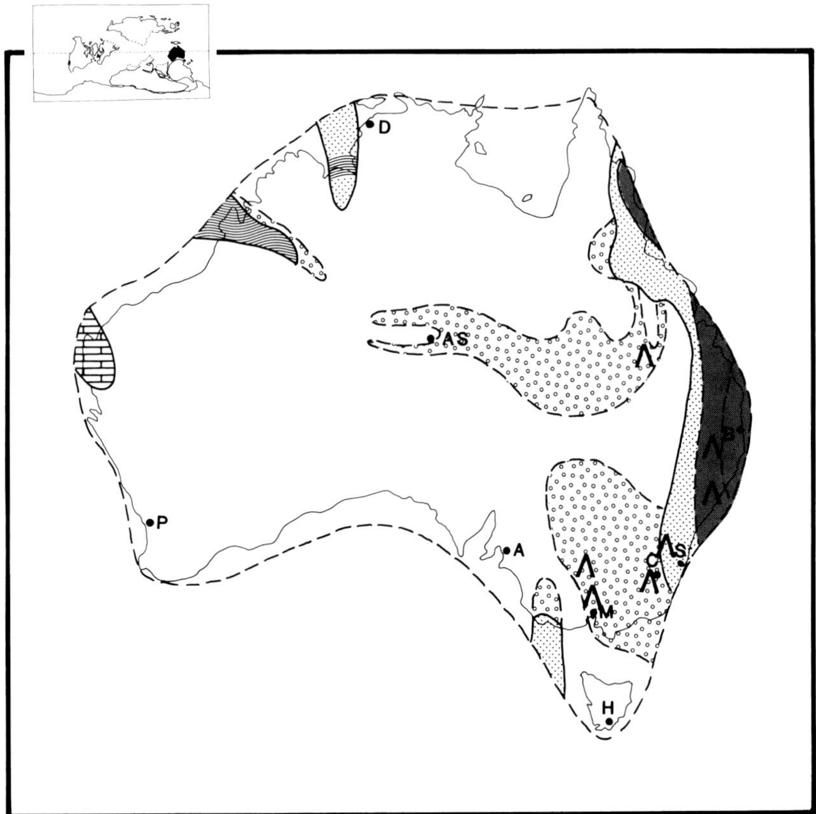

Figure 2.8: The Devonian world and Australia, 408.5 to 362.5 million years ago. Ocean waters only touched the continent along the edges, and armoured fish with jaws lived in streams in south-eastern and eastern Australia as well as on the reefs in the north-west. Australia was still near the equator, although it was just beginning a long journey to the south.

OCEAN AREAS

- Shallow seas
- Shallow seas with limestone deposits
- Shallow seas with reefs (light blue)
- Shallow seas with glaciers
- Deep ocean

LAND AREAS

- Dry land
- River deposits
- Coal swamps
- Glacial deposits
- Volcanic deposits
- Volcanoes

Volcanoes would have been making life difficult for fish again during this period, all along the eastern coast of Australia. Of course, fish and geologists don't always agree about things - what's bad to a fish in this case makes geologists happy! Once they solidified, the rocks that were knocking the fish out, could be dated by studying the radioactive minerals in them, so geologists could tell just how old they were in years before the present (ybp). Because fish fossils were sandwiched in amongst these volcanic rocks, palaeontologists could then tell how old the fish were. A "palaeofishologist" (properly known as a palaeoichthyologist) might point at the latest discovery and say, "This fish is just about 400 million years old, plus or minus 10 million years or so. I wouldn't like that for tea tonight!"

LIFE ATTACKS THE LAND

It was during this period that life began to ooze or crawl out onto land. First, so we think, went the plants - algae to begin with and later, sometime during the late Silurian, true land plants appeared. A true land plant is a plant that can survive on land without having to spend much or any of its time in water. Plants managed this by developing structures that gave them support to stand up on their own - it was about time! Such support was aided by the development of vessels - tubes that transported water and nutrients in the plants - which allowed the plants to become large enough to form bushes and trees. Without vessels it wouldn't have been possible to transport life-sustaining materials (food and water) to distant parts of a plant. Plants that lived on land also had to have some sort of protective armour to keep them from drying out, so a tough outer skin or cuticle developed.

The last problem that had to be solved was how to reproduce other plants. If the plant was not to be the last of the line, offspring had to be produced. Spores seemed to be one way of solving such a problem. Spore-bearing plants, such as ferns, go through two kinds of reproduction in their life cycle - one called sexual (and that's what everyone seems to know about!) and another called asexual. Spores have to do with the asexual kind of reproduction. Spores are carried off by the wind, settle down and develop into male and female fern plants, which are so small that you could step on a thousand of them and never know it. The male and female plants produce cells that are either male or female, and when these get together, there is sexual reproduction. This kind of reproduction has to occur in moist places, so plants with this kind of reproductive plan are never too far away from moist, swampy areas. That means that they would certainly not be living in a howling desert. And these were probably just the kind of plants that were around in the Silurian and Devonian. They lived on the land, but were somewhat restricted in where they could grow. They were not true land plants. Not one of them had seeds, and plants really needed seeds before they could survive in the harsh parts of the terrestrial environment and cut their links with the swamps.

Land animals followed the plants onto land, and Chapter 7 deals with these slithery creatures, the labyrinthodont amphibians that seem to have made this move in many parts of the world sometime in the late part of Silurian times.

The Carboniferous and Permian 362.5 to 245 million years ago

Huge changes began to occur at the end of the Devonian. For quite a long time, about 200 million years, the Australian continent had clung close to the equator and enjoyed a tropical climate. Towards the end of this time, it began moving south, and by the middle and end of the Devonian it was as much as 40 to 50 degrees of latitude further south. That distance is equal to about half the distance between the equator and the north or south pole. That's certainly an impressive distance to move so much rock. This, of course, meant that the climate affecting the continent would also have changed dramatically during this period. If you think that weather forecasters have problems now, imagine being one during the late Palaeozoic.

Australia left behind the tropical warmth, and moved into the polar coolness. As a result, by the late Carboniferous Period, massive glaciers developed, starting out sneakily as mountain glaciers in the volcanic highlands along the east coast, but later developing into thick, continental glaciers that nosed like big bulldozers across many parts of the Australian continent. Other parts of the world had similar problems. Huge glaciers existed in south-western, central, south-eastern and north-western Australia, and they left behind traces, such as polished rock surfaces with big grooves. These were made by gravel and boulders, trapped in the bottom of a thick glacier, which cut into the land surface as the massive ice sheets moved by.

Figure 2.9: The big freeze began during the Carboniferous in Australia, 362.5 to 290 million years ago. Gigantic glaciers pushed across several parts of the continent, leaving behind debris and polished rock surfaces, some with deep grooves scratched by rocks trapped in the undersides of glaciers. Australia now lay far south, after moving from its place near the equator in the Devonian. It would stay at these high latitudes, a next door neighbour to Antarctica for nearly 250 million years.

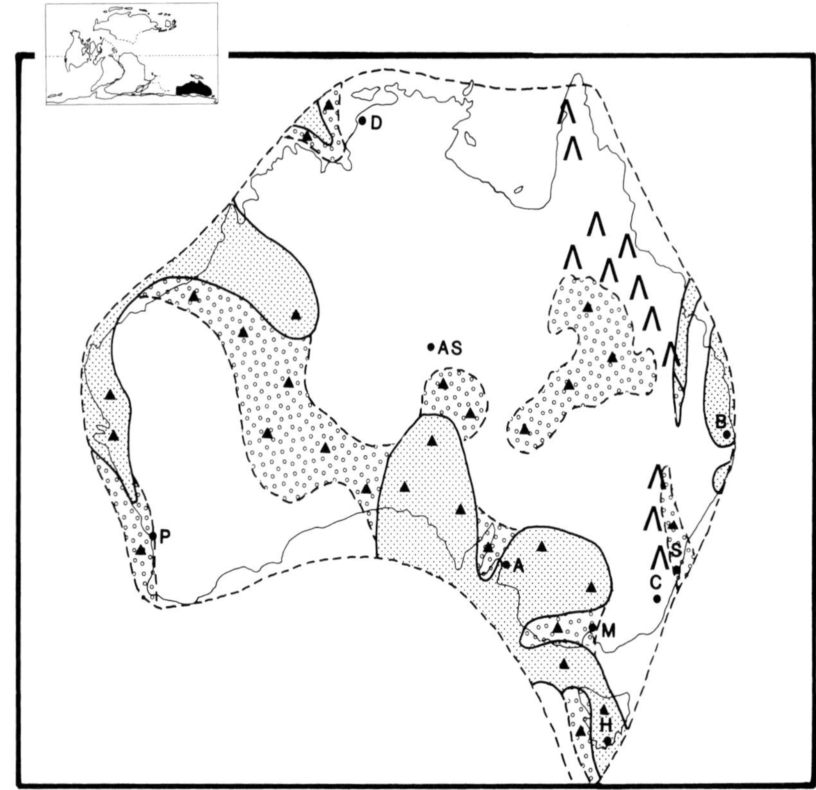

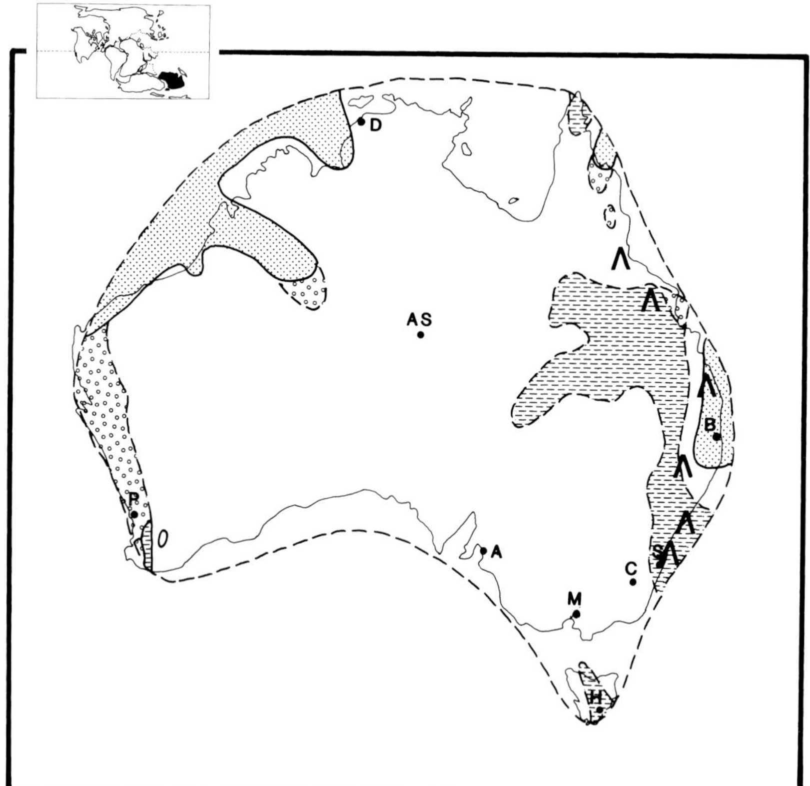

Figure 2.10: Sometime during the Permian Period (290 to 245 million years ago), the climates began to warm up. Massive swamps and bogs formed along Australia's east coast at this time, and in these were formed the beginnings of our black coal deposits, such as those now mined at Newcastle, NSW and in Queensland.

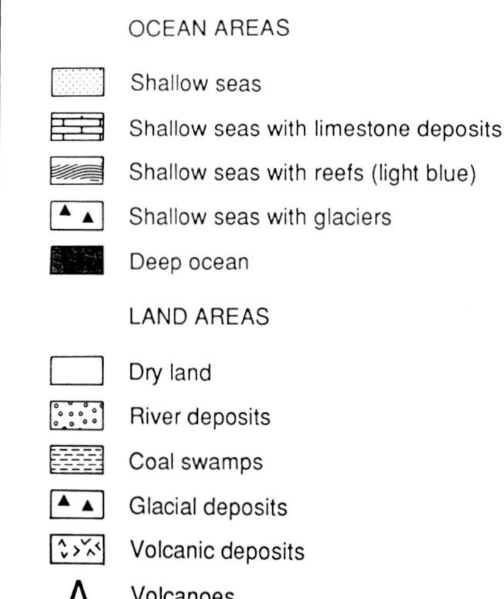

Towards the end of the Permian, things began to warm up, the glaciers melted and sea levels rose. Why did sea levels rise? There is only a certain amount of water present on the Earth's surface. If much of this is trapped in ice sheets or glaciers, then the sea level is lower, as it is now. Just the opposite happens when there are no ice sheets at all - the sea level is higher. If the polar ice sheets of today melted right now, and some climatologists (people who study climate) are worried about this, then cities like Melbourne, Sydney and Brisbane would find themselves underwater, because of the consequent rise in sea level. And that's not too terrific for those of us who have houses in the low-lying areas!

On the low ground during the Permian, large areas of swamp land in such places as Newcastle and southern Queensland gave birth to coals - a major source of power in our country today. In the Permian you would have been up to your knees or waist in slimy, soggy mud in places where today massive dredges cut through solid coal. That just shows you what a couple of hundred million years of burial can do!

By the end of the Permian, the Australian climate was not frigid, but cool and temperate and generally fairly wet. The stage was set for the Mesozoic Era, the time reserved for the dinosaurs and their friends.

A WALK ON A MESOZOIC TERRAIN: IN THE FOOTSTEPS OF THE DINOSAURS

The Triassic
245 to 208 million years ago

During the Triassic you would have found walking in Australia more productive than swimming, unless you preferred rivers to sea water. Marine realms existed mainly in the far north-east and north-west. Australia didn't move much either, during this time - it stayed far south in the cool, humid latitudes and close to the other parts of Gondwana. Gondwana was joined to the northern continents at this time too, so you probably could have walked to Africa or to America, as long as you had good shoes and an iron will, and kept close look-out for hungry, mammal-like reptiles. This wide open highway to all the continents was not to last forever, however.

OCEAN AREAS

- Shallow seas
- Shallow seas with limestone deposits
- Shallow seas with reefs (light blue)
- Shallow seas with glaciers
- Deep ocean

LAND AREAS

- Dry land
- River deposits
- Coal swamps
- Glacial deposits
- Volcanic deposits
- Volcanoes

Figure 2.11. During the Triassic, 245 to 208 million years ago, Australia was mostly above sea level, and river and lake deposits were forming in many places. The continent still lay far south and was part of the great southern land area called Gondwana.

Big rivers left their sands and gravels behind in many parts of Australia at this time. You could have paddled right around Sydney on a river as big as the the Amazon or the Congo. The last remnant of this giant watercourse is the Hawkesbury Sandstone that forms the cliffs around Sydney and out of which many of Sydney's buildings are constructed. Further north in Queensland, rocks of similar age were laid down, and the muds and sands of these rivers formed the graves for many different kinds of crocodile-like amphibians.

The Jurassic and Cretaceous 208 to 65 million years ago

Australia had been quiet for a long time, but at the beginning of the Jurassic all of that changed. Gondwana was beginning to break up, and enormous outpourings of lava and related volcanic rocks resulted. These rocks can be seen in Tasmania and Antarctica and on many other parts of Gondwana. They tell geologists that the supercontinent was splitting apart.

If you could have taken a walk around the continent in the Jurassic, you would have found it much the same as it had been in the Triassic. There was very little in the way of oceans except in the north-west, and a very small area of sea just north of Brisbane. Just as in the Triassic, there were active volcanoes along the eastern coast, in addition to some new volcanoes along the south coast, where Australia was beginning to break off from Antarctica. There were big rivers all over the continent, and a few swamps and lakes as well, but the lay of the land would have been familiar to you from times past.

You would have been totally disoriented, however, just a little later in the Cretaceous. The continent was invaded by shallow sea waters. The same sort of thing happened in many different parts of the world, and we are not sure why. Australia was still connected to Antarctica, but the sea invasion changed Australia from a large land mass to a group of islands, as the palaeogeographic map in Figure 2.12 clearly shows. Then Australia got serious about its divorce from Antarctica, and a deep fault valley formed along the south coast. The two

Figure 2.12: Sea monsters - plesiosaurs - quite enjoyed themselves in Cretaceous Australia, 145.6 to 65 million years ago. All over the world including Australia, shallow seas flooded onto the continents. Australia was cut into at least four separate islands. This was a time when climates were really warm over much of the world, perhaps as much as 17degrees Celsius above present temperatures. This great warmth may have been partly the cause of the flooding of continents, because it melted all of the ice that probably lay at the poles. This flooding may also have been caused by the increased rate of the spreading of the sea floor, which would have added heaps of volcanic rock along the ocean ridges and reduced the capacity of the ocean basins. As there was less room in the ocean basins, the water had to go somewhere, possibly onto the land. With all that water on the continents, plesiosaurs were abundant in central Australia. Their opalised skeletons have been preserved in rocks of the Cretaceous age in such places as Andamooka in South Australia and White Cliffs in New South Wales.

continents were beginning to move apart as though a zipper were opening from the west, and the valley thus formed between the two emerging continents was the forerunner of a proper ocean basin, not too far off in the future. Into that valley poured mountain streams, bringing ice-cold water and a heavy load of sediments from the snow-covered mountains that perhaps lay off to the north, south and perhaps east. Those sediments were laid down in the ever deepening valley, burying carcasses of dinosaurs and fish that lived in the valley. This part of Australia was far south at the time, as much as 85 degrees, and so it was well inside the Antarctic Circle. This meant that the dinosaurs and their friends would have had to survive a three-month-long winter night. Perhaps 110 million years ago it would have been a lot easier than today, because it seems that the whole world was a great deal warmer at the time and there were no major ice caps at the poles, as far as we know. So, maybe the dinosaurs could have made it through the long winter night without much stress, perhaps by hibernating, even migrating, or maybe even just by plodding on through the darkness.

Towards the end of the Cretaceous, temperatures began to cool once again. The shallow oceans that had covered much of Australia drained off the continent - this was probably somehow related to the build-up of ice at the poles. Extensive river systems again developed where before huge sea monsters, the plesiosaurs, had swum, and now dinosaurs left their footprints on the broad floodplains - the best-known examples of these were found in the area around Winton in central Queensland. Soon, however, this was all to change drastically. Australia was already embarking on another, incredible journey.

The Tertiary Period
65 to 1.64 million years ago

A TREK INTO THE CAINOZOIC RAINFOREST DESERT

Most people probably think that Australian deserts are old, perhaps as old as the Australian continent itself. In fact, our deserts are really quite young. These deserts, such as the Simpson Desert, in the very heart of the continent, are great places to go fossil-hunting. The rocks are laid bare, and geologists go crazy in this country, checking out all the outcrops (that is, all the rocks that are jutting out here and there). These very rocks that you see as you trek across the desert wastes north of Marree in South Australia are actually the remains of ancient lakes and streams that once lay and ran upon a gently rolling and humid land some 15 to 20 million years ago, during the Tertiary Period. As you feel the grit of sand between your teeth and the flies crawl into your eyes, it's hard to believe that platypuses and dolphins once played in the watercourses here not so long ago.

Through most of the Tertiary Period much of Australia was wetter than it is now or has been during the last 100,000 to 200,000 years. Rainforests flourished in the centre of the continent, and big rivers wandered across the land, as the maps for the early and middle Tertiary clearly show (Figures 2.13 and 2.14).

At the same time, Australia itself was on the move. You may remember that we mentioned that great volcanic outpourings in the Jurassic were the warnings of things to come: the event they foretold came in the early Tertiary when Australia finally

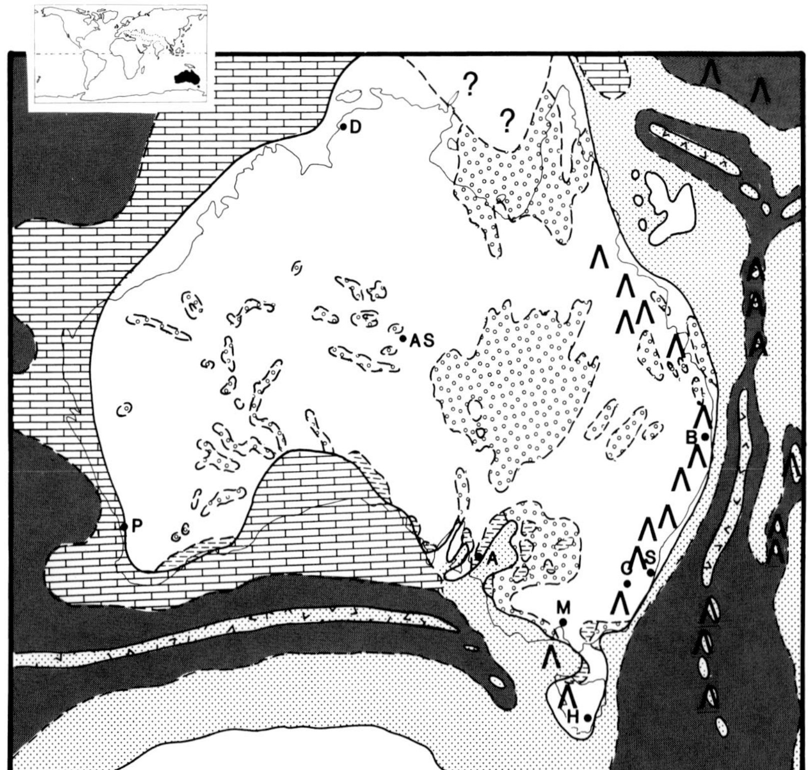

Figure 2.13: The early part of the Tertiary, 65 to 23.3 million years ago was relatively wet in Australia. River and lake deposits are known in many parts of the country. Pollen and spores in these rocks tell us that rainforest was abundant over much of the continent, right into what is now the dead heart of central Australia. All that was to change later.

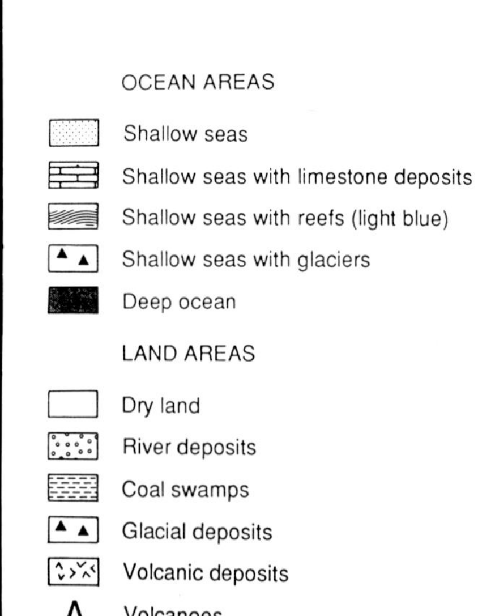

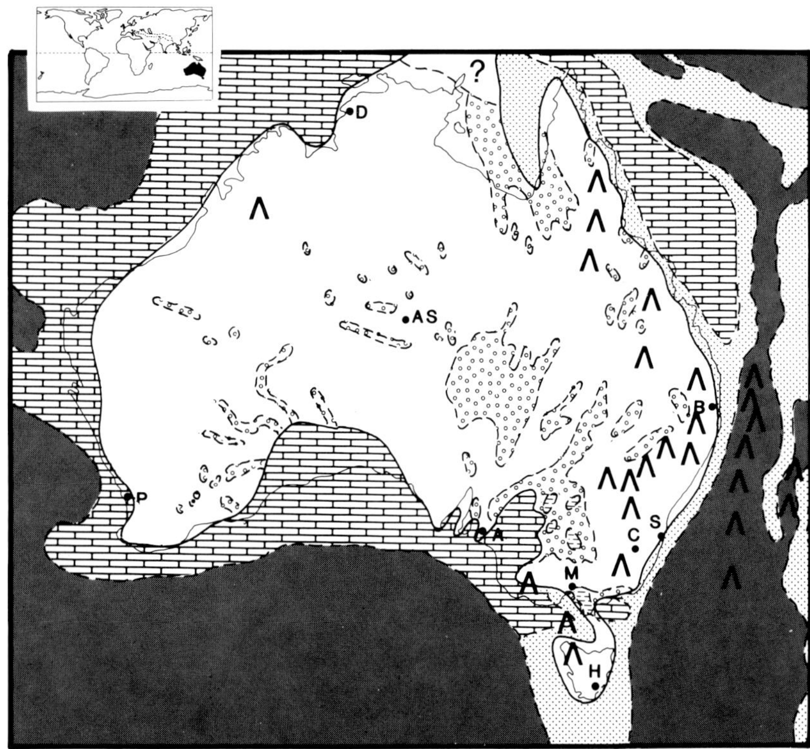

Figure 2.14: The last 23.3 million years (the middle Tertiary to the late Quaternary) on this continent has been marked by great change. Australia broke away from Antarctica and sped north towards the tropics, from where it had come millions of years before. It was during the last 23 million years that Australia gained its tropical animals and plants, and during the last five million years its arid centre developed, with rainforests giving way to *Spinifex* grasslands. Animals and plants, like rodents and wattle and most recently humans, successfully migrated from the north to our long-isolated continent. Its more ancient inhabitants, such as its marsupials and waratahs, had been born on a great southern landmass, Gondwana, or on the isolated Australian continent itself.

said goodbye to Antarctica. About 45 million years ago, a seaway broke through from the west to separate Australia from the rest of Gondwana, and from then on Australia was on its own - a lonely ship sailing northwards toward China. Some people have called it a "Noah's Ark, " for it carried with it all the animals and plants that it had once shared with Antarctica and South America. These animals and plants then became quite separated from the rest of the world from that time until the present. This is one of the main reasons we have such an interesting collection of native animals here compared to those elsewhere in the world.

SAND AND ICE

The Quaternary Period
1.64 million years ago to the present

In our trek through the Cainozoic, we would have seen rainforests, rainforests and more rainforests (though not tropical until recently), with an occasional grassy area here and there, but not on the scale of the vast plains we are used to today. Once we step into the Quaternary (and to be truthful just a little before this) things begin to change, at times rather quickly. If we had looked *far* to the south, we would have seen the Antarctic ice cap growing. It had begun to grow earlier, in the Tertiary, but now it was really becoming enormous. In fact, at times during the Quaternary it probably grew larger than now. In a nutshell, the world, and Australia in particular, began to look like it does today. Central Australian lakes dried up and not so long ago,

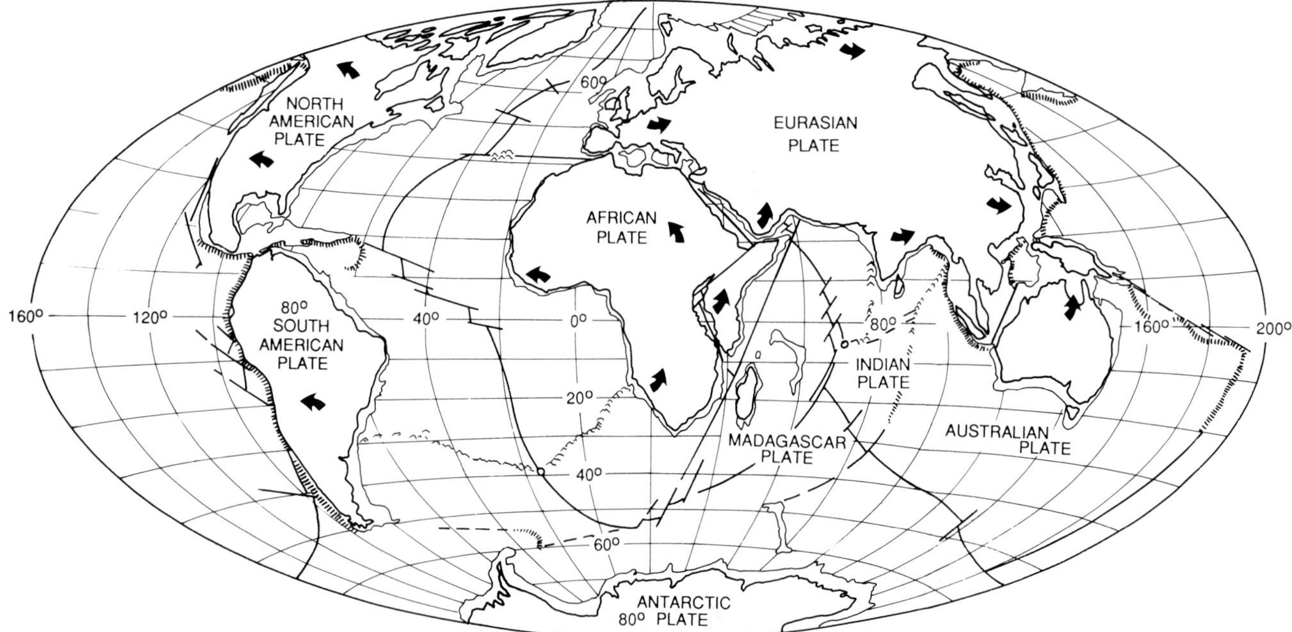

say about 20,000 years, Australia was even drier than it is now - these were the "prune years" - dune fields in South Australia, the Northern Territory and Western Australia were on the move, and blowing sand would have stung your face. Today, grass covers several of these big dune fields in many places because the climate is wetter now and encourages plant growth. If there were less rains, the dunes would be on the move again.

At times during the Quaternary, huge glaciers moved southwards from the Arctic to cover much of North America, Europe and Asia. Australia had no big glaciers like this; there were only mountain glaciers here. Those glaciers in the northern hemisphere did affect our climate, however, for during the great glacial advances, sea levels were lower worldwide. These were times when land bridges connected Australia to both New Guinea and Tasmania. These were also the driest times for Australia.

Figure 2.15: What will the future bring? This map is a prediction of what the world will look like in 50 million years. Once again, Australia is not a separate continent, but a part of Asia - Australasia would be a fitting name. How the plants and animals will greet one another, and for that matter which ones will still exist, will depend on what you and I and our decendants do to this world over the next few million generations. Will life still exist then? Or will Earth's climate be like that of Venus or Mars, too hot or too cold for the life we know?

CHAPTER 3:
ONE-CELL BLOBS

Single Cells, the Origin of Life, the First Animals and Plants and Their Multitude of Cells

---

The Precambrian
4500 to 1590 million years ago

The rocksaw hummed in the corner, water spraying the plastic shield around it as the corundum studded blade slowly ate through a piece of glassy rock, a chert, from the MacDonnell Ranges. Ian flipped off the switch as the saw finished its job. Then he took the tiny 5 centimetre slab of rock and promptly glued it to a small glass slide. He then began an even longer job, the slow polishing of that thin slab of rock to an even thinner bit. Ian put an abrasive powder on a big, wet piece of thick glass and turned the slide upside down after the glue had set to hold the rock firmly to the slide. His hands moved the inverted slide round and round. The powder gradually wore away microscopic bits of the rock until, when Ian held the slide and its attached rock up to the light, he could see right thorough it. His "thin section" was ready now. He washed it off more throughly, dried it, then put it under a reasonably powerful binocular microscope, and gazed down the two barrels at remains of a 900-million-year-old bit of sea bottom. There they were, about 10 little circles, minute even at this magnification of nearly 500 times - the fossilised remains of some very old single-celled life forms, probably algae.

Only now did Ian know that his trip to Alice Springs had been a success! He had journeyed over some rather questionable roads, even through countryside that had never known roads, and he had collected several hundred kilograms of rock samples. Once he got back in the lab he had spent hours preparing the samples. Only now did he know he had successfully collected fossils! All of his work would have been in vain if those little cells had not turned up after hours of tedious lab work. Ian went off for a tea break with a smile on his face, wearing a dirty white lab coat that was decorated with the remains of the fossiliferous chert.

Australia has the oldest known rocks in the world. These are more than 4,000 million years old. In fact, trying to think about such great chunks of time is rather mind boggling. You have to try to relate it to something familiar, such as perhaps the length of a human life. For example, you might be lucky enough to live to 90 years old. So as to make a comparison with the time scale of the Earth's life, let's say the Earth formed and started to cool when you were born. The only difference is that your birth was perhaps 12 or 15 years ago, but the Earth's birth was at least 4,500 million years ago. On that scale, when you were about one and a half years old, the first signs of life began to appear - single-celled organisms that could reproduce (in Earth time that was about 3,800 million years ago). When you are about 70 years old, the life forms that were preserved in the rocks Ian was cutting up from the MacDonnell Ranges were living (some 900 million years ago). By your 78th birthday, the first living cells that had a centre or nucleus where they concentrated their information about inheritance were in existence. Some of these life forms also had more than one cell - they were multicellular, and they have left abundant fossils in the Flinders Ranges of South Australia, where they lived and died about 600 million years ago, Earth time. On your 86th birthday the dinosaurs were your best present. They appeared about 200 million years ago, Earth time. Warm-blooded, hairy mammals got their start about the same time. But the group to which we belong, the primates, first appeared on the scene just before your 89th birthday, and our species appeared in the very

Figure 3.1: The Precambrian seas were probably the cradle of life, but it was a long time from the appearance of crude single-cells to the many celled life forms that are first known as fossils only 600 million years ago. (From the *The Tower of Time*, a mural by John Gurche, courtesy of the Smithsonian Institution, Washington, DC)

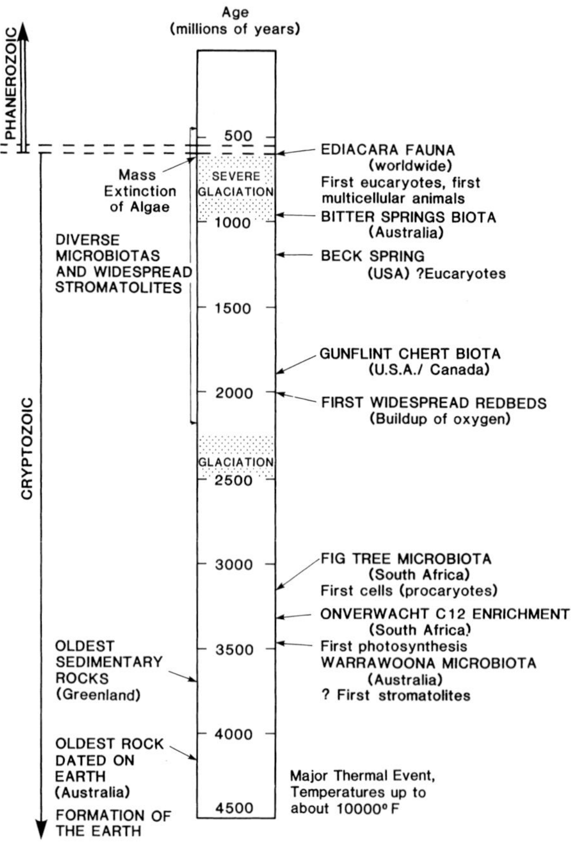

Figure 3.2: A Calendar of events in the Precambrian from 4500 to 600 million years ago. Although the first cells are shown here as occuring about 3200 million years ago, there may be some in Western Australia that are as old as 3400 to 3500 million years, maybe even older!

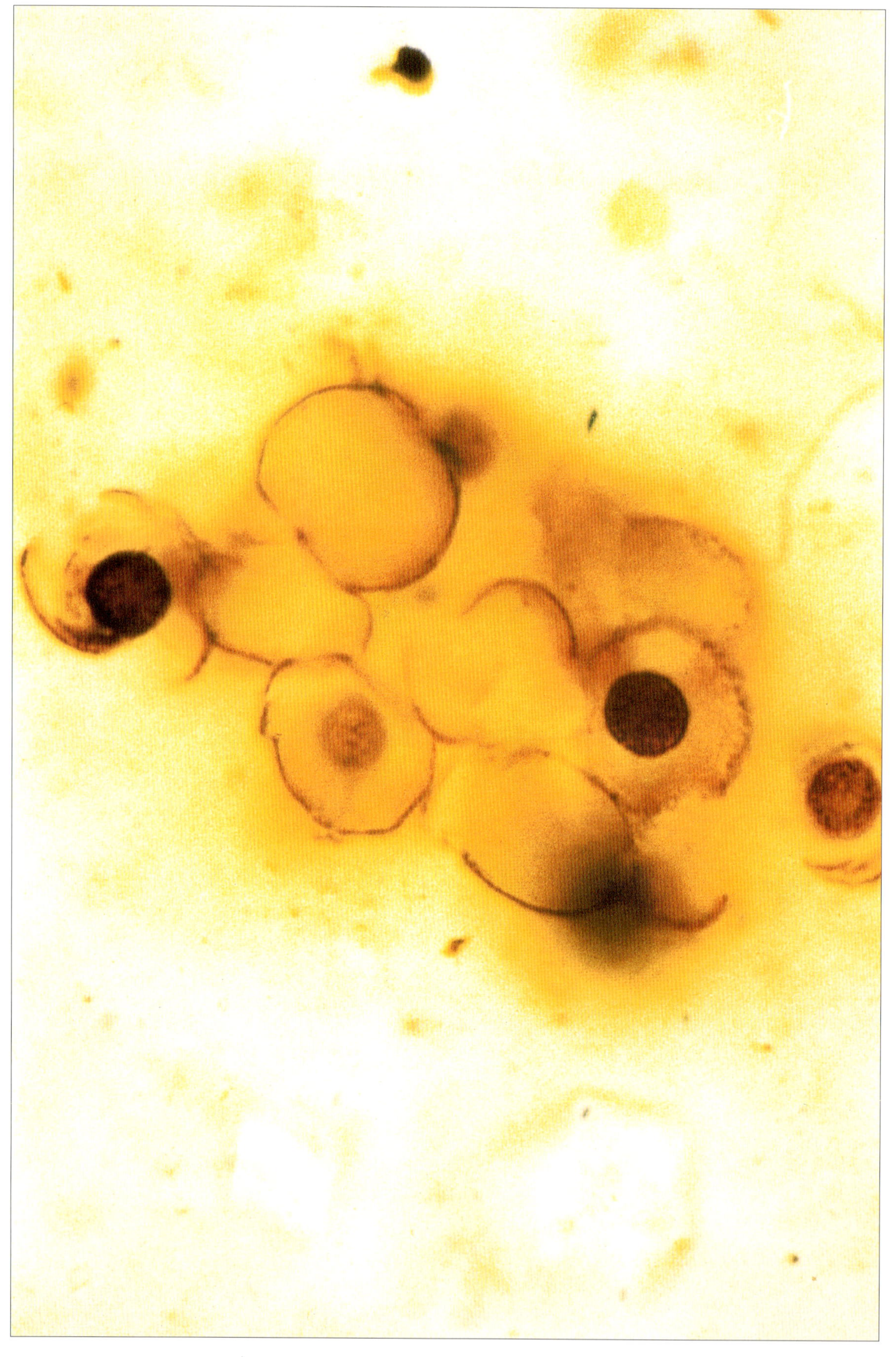

Figure 3.3: (*Facing page*) Single-cell life forms preserved in shallow ocean rocks about 700 million years old from the MacDonnell Ranges, Northern Territory. These cells are only about one-hundredth of a millimetre across, invisible to the naked eye. Even though these cells seem to have a central nucleus, the nucleus is actually a mineral particle. These tiny cells are probably the remains of bacteria or algae. (I. Stewart, J. Warren)

THE BEGINNINGS OF LIFE

last days of your life. Our genus, *Homo*, appeared between two and three million years ago, and our species, *Homo sapiens*, sometime in the last few tens of thousands of years, Earth time.

Returning to the past - the far distant past in Australia - we'll look in this chapter at the origin of life and its history during the first 4,000 million years of Earth time, in terms of a human lifetime, up until you have your 80th birthday - really quite a span of time!

In the north-west of Western Australia you can collect some rocks rather similar to the cherts in the MacDonnell Ranges, which up to now have yielded some of the oldest remains of life on Earth, the Warrawoona Microbiota (microbiota refers to microscopic life forms). Fossilised cells of these life forms and the mounds they built (stromatolites) have both been found. Interestingly, if you want to wade or swim through an environment just like that of the MacDonnell Ranges or north-western Australia 3700 to 3800 million years ago in the MacDonnell Ranges or in northwestern Australia, you will find it at Shark Bay on the west coast of Western Australia, south of Carnarvon. There, algae and bacteria form glue-like mats that trap silt and clay particles, which in turn build mounds that grow up to the high tide level. These are living stromatolites, playing out their lives probably much as their ancestors did thousands of millions of years ago. Today, stromatolites live in only a few places of the world, such as the very salty bays and lagoons of Western Australia and the Persian Gulf. This was not true in Precambrian times, however. Stromatolites then occurred all over the world. In the Precambrian there were few, if any, predators that attacked them. Once sea-dwelling snails came on the scene - about 600 million years ago - the stromatolites were under attack and rapidly retreated.

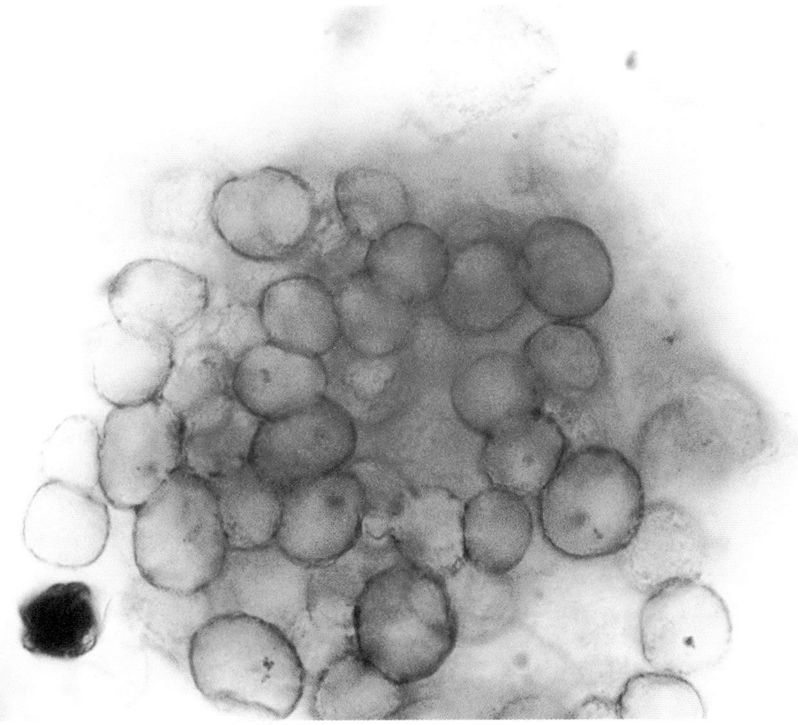

Figure 3.4: A mass of fossilised single-celled organisms from the Bitter Springs Formation (Precambrian) of the Northern Territory (I. Stewart, J. Warren).

Figure 3.5: Fossil stromatolites from Precambrian rocks (Proterozoic Eon) in the MacDonnell Ranges, Northern Territory, west of Alice Springs. For billions of years, the largest structures built by living things were these. Stromatolites are the remains of sediment trapped in a glue-like sheet formed by these one-celled algae and bacteria, which lived on the edges of the shallow seas. The black-and-white scale in this drawing is divided into centimetre squares. (J. Warren, I. Stewart)

Besides the fossilised cells and stromatolites, the iron-rich rocks from Western Australia may also be signs of ancient life. These Banded Iron Formations (BIFs) may reflect times when there were varying amounts of oxygen in the environment. The red bands may indicate the presence of plenty of oxygen and the dark bands, times when there was little oxygen.

Oxygen was a big issue in the Precambrian. In fact, you wouldn't have been able to catch your breath until very late in this time period. It was the activity of living things that made the oxygen we breathe today - their gift to us. For the first 2,000 million years of the Earth's history there wasn't much oxygen at all.

How did life get started? That's a question that lots of people have different answers to. Looking at the information we have at hand from laboratory experiments and the rock record, one generally popular theory goes like this:

Put some energy into the kind of atmosphere that was hanging around the early Earth, most likely made by volcanic eruptions. That atmosphere included water vapour and either hydrogen, ammonia and methane (swamp gas), or nitrogen, carbon dioxide (the gas plants generate at night) and carbon monoxide (the fumes that come out of car tailpipes and are dangerous to breathe); when you zap this kind of primitive atmosphere with lightning bolts, thunder, intense sunlight, or freezing temperatures, the building blocks of living proteins are produced together with those of DNA - the central substance that codes our hair colour or face shape and other characteristics. DNA, or deoxyribonucleic acid, is the molecule containing the basic genetic information of all living things.

Figure 3.6: Small stromatolite columns preserved in a limestone rock, which has been cut in half. The layering in the columns shows how one colony has built on top of another.

Figure 3.7 (a): If you ever want to walk back in time 2,000 or 3,000 million years ago, go to Shark Bay, Western Australia. Here, stromatolites are growing today as they did in many parts of the world thousands of millions of years ago. These stromatolites are lined up parallel to the ingoing and outgoing tidal currents, so they can be used to tell which way the water is flowing. (M. Fenton)

Figure 3.7 (b): Stromatolites at Shark Bay, Western Australia living in an area battered by strong waves. Stromatolites tend to grow in circular masses under these conditions. (M. Fenton)

(b)

(a)

This may sound a bit like a fairy story - but in fact, experiments have been carried out in several biochemical laboratories in the world where substances such as amino acids and DNA building blocks (nucleotides) have been produced from non-living, inorganic molecules that exist in our environment today, and also existed in the Precambrian. So far, the final step to go from building blocks to the DNA molecule, to a structure that can reproduce itself - has not been accomplished. But biochemists are working on it - their science, like the science of computers, is still in its infancy. Some work now shows that when laboratory produced building-blocks (molecules) are mixed in water, they cling together in small, sphere-like structures separated from the water by a thin skin or membrane. Maybe this will eventually lead us to understand how reproduction began and cells first formed. The fossil record just shows when cells appeared and the history of life after that. It's up to laboratory experimentation to show us how we get from building blocks to duplicating cells.

THE FIRST MANY-CELLED ORGANISMS

Probably the most important fossil site in Australia is in the Flinders Ranges in South Australia. Imagine your surprise if, when working in the dry, saltbush covered hills near Wilpena Pound, you came nose to nose with a jellyfish - or a worm solidly stuck in a rock? That is exactly what happened to Reg Sprigg when he was a student at the University of Adelaide, and some of his professors told him he was seeing things - but he decided he wasn't. Reg had discovered some wonderful animals - the oldest life forms with more than one cell. Now called the Ediacara fossils, they include many different kinds of animals, such as sea pens, perhaps ancient relatives of the trilobites (see Chapter 4), but mostly worms and jellyfish. None of these animals had skeletons - it's a wonder that their soft bodies left delicate and detailed imprints in the beach sands over 600 million years ago.

Figure 3.8: A Banded Iron Formation (BIF), perhaps the result of greater and lesser amounts of oxygen available in the Precambrian seas. Oxygen at this time was probably produced mainly by living single-celled forms. (S. Morton)

Figure 3.9: (*Right above*) A fossil jellyfish from Wilpena Pound, South Australia, in Precambrian rocks about 600 million years old. These jellyfish grew up to several centimetres across and are some of the oldest animals known. (N. Pledge)

Figure 3.10: (*Right below*) A fossil marine worm from the Ediacara fauna found in Precambrian rocks at Wilpena Pound, Flinders Ranges, in South Australia. These fossils are preserved as imprints of the original animals. The actual animals disappeared long ago. The carcasses of these ancient animals may have been washed up on an ancient beach (which formed the Pound Quartzite) and preserved there. Although the first large collections of Ediacara animals were found in Australia, they are now known from many parts of the world. (N. Pledge)

CHAPTER 4:
ICE AND EXPLOSIONS

Ice Ages, Many More Cells, An Explosion of Armoured Life

The Late Precambrian and Earliest Cambrian 900 to 500 million years ago

We sat on top of a low ridge off to the east of the main part of the Flinders Ranges. Leaellyn had been sliding down a big sand dune that hung over the ridge, leaving long grooves on its steep slope. The sand that made this terrific slide was really quite young - that is, when compared to the ridge it lay on. We had been poking about along the ridge all afternoon and had dug out many fossils. Some looked like the corals you might find on the Great Barrier Reef today if you put on flippers and a snorkel. But were they really all that alike? A closer look told me that there were some real differences. These ancient sea creatures had an extra something! In living corals there is an outer wall with lots of little walls or partitions running away from that wall at about 90 degrees - and that's that. In my fossils there was also an outer wall and partitions, but the partitions were connected to an inner wall. So, my "corals" were really not *corals at all, but tricky look-alikes. They built reefs just as corals do, but they were only pretenders!*

The ridge we were on was actually the remains of an ancient reef that had grown in a tropical sea, now drained from the land. Over millions of years that shallow ocean had turned into a desert. The sand dune that Leaellyn had been sliding on had formed in that desert very recently.

Mixed in with the coral look-alikes, archaeocyathids, were the clam-like brachiopods and occasionally a trilobite, a now extinct, sea-dwelling creature that was a distant relative such arthropods or joint-legged animals as slaters, crabs and cray-fish.

The Flinders Ranges off in the distance contain the Ediacara fossil jellyfish and worms found by Reg Sprigg. The animals on our ridge are quite different from those that Reg found. Ours lived a few tens of millions years later in time. They are also different in another very important way from the Precambrian Ediacara creatures. Our archaeocyathids have armour: they have skeletons, which were lacking in Reg's animals. Something very important had happened between the

Figure 4.1. Archaeocyathids were abundant and widespread in the early part of the Cambrian and were extinct by the end of that period, 505 million years ago. They were probably an early experimental model that lost out when the true corals got going, but were successful for several million years before corals appeared (S. Morton).

THE BEGINNING OF HARD SKELETONS AND A GOOD FOSSIL RECORD

Figure 4.2. Wilpena Pound in the Flinders Ranges of South Australia is a place to hunt the first animals with many cells. Here rocks about 600 million years old preserve the fossils of jellyfish, worms and sea pens (related to starfish), as well as several animals that have no known relatives today. (N. Pledge)

time the Ediacara animals lived and the early Cambrian time when our archaeocyathids built their reefs in many parts of Australia, when Australia lay very close to the equator and far away from where it is today.

Figure 4.3. South rim of Wilpena Pound in the Flinders Ranges, where the Pound Quartzite, a hardened, silica-rich sand, has preserved 600 million-year-old shallow water marine life. (N. Pledge)

Figure 4.4. *Dickinsonia costata*, a worm-like animal that was about 6.5 centimetres long from the Precambrian rocks of Wilpena Pound. (N. Pledge)

Figure 4.5. *Parvancornia minchami*, an animal from the Precambrian rocks in Wilpena Pound, whose relatives are not known, was about 2 centimetres long. (N. Pledge)

Figure 4.6. *Mawsonites spriggi*, a jellyfish-like animal of Precambrian times named after Reg Sprigg, who discovered the Ediacara animals, and Sir Douglas Mawson, the Antarctic explorer. Fossil found at Wilpena Pound. (N. Pledge)

Figure 4.7. *Tribrachidium heraldicum*, a Precambrian animal whose relatives are unknown, was about 2.5 centimetres across. Fossil found at Wilpena Pound. (N. Pledge)

Figure 4.8. A reconstruction of some of the Ediacara animals by Robert Allen and Mary Wade. (Courtesy of the Queensland Museum) These animals lived over 600 million years ago in Australia.

Figure 4.9. In Cambrian times, Australia lay near the equator, and many parts of the continent were covered with shallow oceans. Shallow banks in these warm tropical seas, like those pictured here, were perfect spots for reef- building animals, such as archaeocyathids and algae, to live and build their homes. Quite large reefs of this age can be seen in the Flinders Ranges of South Australia.

The late Precambrian from about 1,000 million years ago to about 600 million years ago, was a pretty exciting time. Imagine sitting on your continent right on the equator with a huge glacier moving slowly over you. Glaciers, small ones, are active today on the tops of some tropical mountains, but nowhere except in Antarctica and the Arctic are there massive continental glaciers - certainly not in Australia.

In the late Precambrian, glaciers formed on almost every continent. It was very cold *everywhere,* even at the equator! This meant that the air, the oceans, *everything* was cold - the coldest the earth had been in a long, long time, perhaps the coldest it had been since before it began to condense out of the blackness of space.

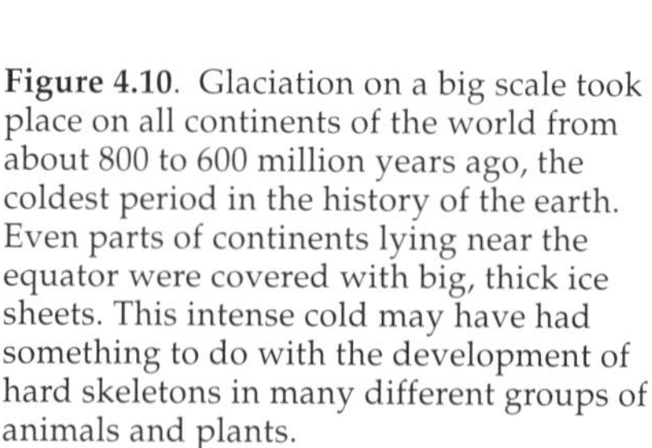

Figure 4.10. Glaciation on a big scale took place on all continents of the world from about 800 to 600 million years ago, the coldest period in the history of the earth. Even parts of continents lying near the equator were covered with big, thick ice sheets. This intense cold may have had something to do with the development of hard skeletons in many different groups of animals and plants.

ICE AGES AND CHANGE

John Shergold, a geologist who works at the Bureau of Mineral Resources in Canberra, thinks that this time of the *big cold* may have had something to do with the armour plating developed by lots of life forms near the end of the Precambrian. How can cold encourage the growth of armour? Armour is not like a fur coat: it can't keep things warm. John Shergold thinks that it might have developed because certain building-blocks needed for constructing armour were made available by the cold. These building-blocks which had been rare before this time were the two elements phosphorous (found in matches) and oxygen, combined into a compound of one part of phosphorous to four parts oxygen (PO_4) - a phosphate. This sort of a chemical compound is common in the deep ocean, but that's not where most things live. John thinks that because it was so cold in the late Precambrian, the water near the surface of the oceans also got terribly cold, and so became dense and sank. As the surface water sank, it pushed up the phosphate-rich water from deep ocean basins, which bathed the animals and plants living near the surface. Some living things must have been able to use this phosphate, first perhaps only as a storage of phosphate used for such functions as energy production, which for instance is used to make muscles contract. Later, phosphates were used to form hard tissues. As a result of this radical change, some plants and animals produced the first hard skeletons from phosphate and calcium (which forms the hard part of our bones and teeth), and so near the end of the coldest part of the Precambrian a number of animals and plants developed skeletons.

Skeletons, especially those on the outside, like clams' shells, crabs' shells, worms' tubes, give great protection. It's harder for another animal to eat those that wear a coat of armour!

Figure 4.11. Trilobites were one of the first groups of animals with protective skeletons to appear in the Cambrian. They were called trilobites because of the three distinct lobes or parts that ran the length of their bodies. Most had complicated eyes made up of hundreds of small lenses, and some may have had three-dimensional vision like humans (S. Morton).

Skeletons also provide a place to attach muscles, to give better control of the movement of jaws and legs. Skeletons give their owners many advantages over their unarmoured cousins. So, in the Early Cambrian, skeletons were the fashion.

For the fossil-hunter, skeletons were the best thing that had happened since the origin of life. Skeletons are much more likely to be preserved as fossils than soft tissue such as skin, guts and eyeballs. In rocks of the early Cambrian Period, palaeontologists (people who study fossils) see the first good record of life preserved - and for the next 600 million years fossils are quite abundant. The explosion of new life forms that happened at the Precambrian - Cambrian boundary has continued right up to now, and it may all be due to the substance that makes a match catch fire - phosphorus!

Figure 4.12. A record of living things showing when their record begins and finishes. The black indicates when their record is best, and this is when geologists can best use these fossil groups to work out the age of rocks. Individual species of animals or plants seem to have a distinct lifetime, for the species. That life span is determined by a great deal of study of fossils over a large area. Then when geologists find the same species in a new area, they can figure out how old the rocks containing the fossils are. If you found an archaeocyathid, you would know that the rocks where you found it were no younger than middle Cambrian and no older than early Cambrian. You would know that they couldn't be Cretaceous or Tertiary rocks. So, fossils act like a palaeontological wristwatch, helping us to measure prehistoric time.

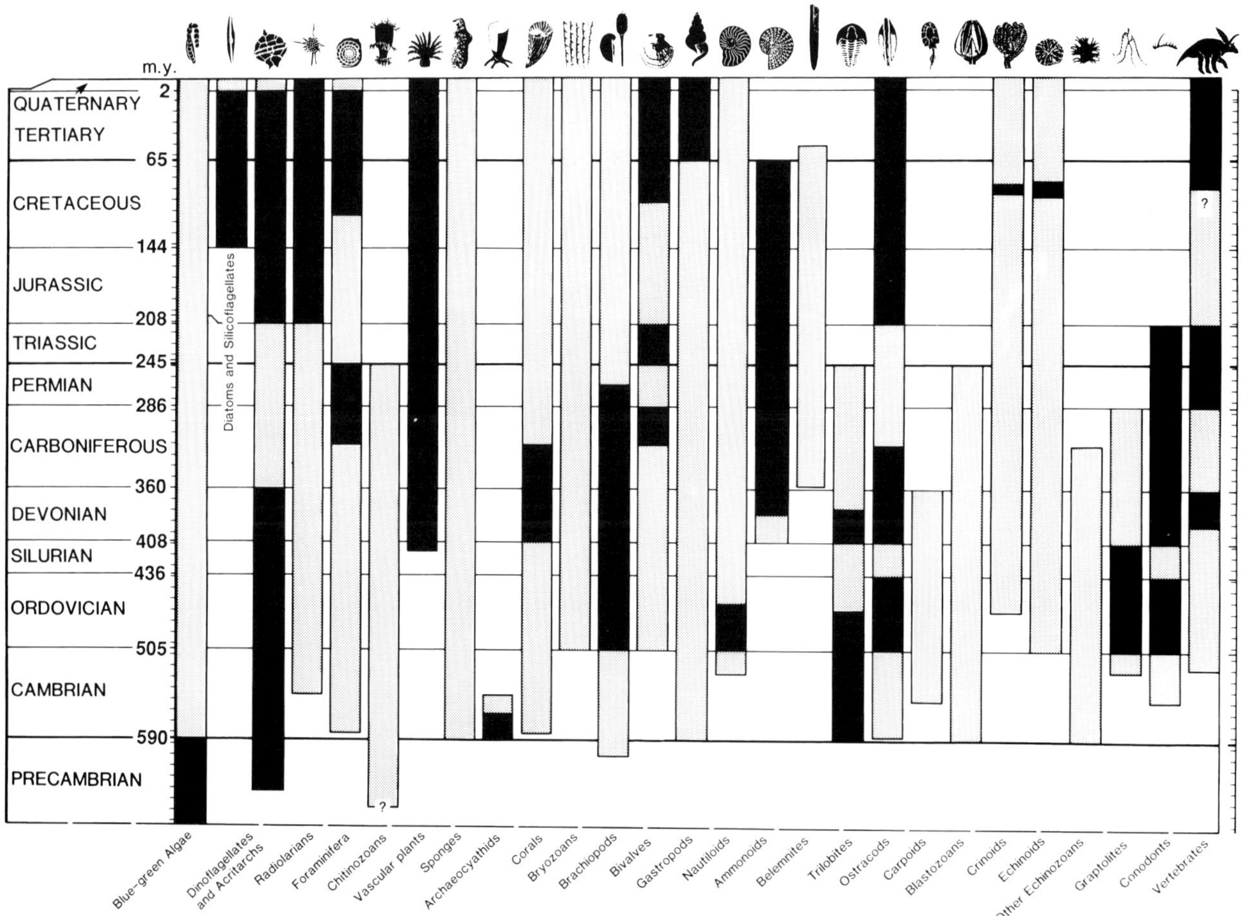

Figure 4.13. Big changes occurred at the end of the Precambrian. Living forms became able to build hard parts and immediately began to leave a much better fossil record. The sponge-like archaeocyathids (below), trilobites (above) and brachiopods (left below) were some of the most abundant early animals with skeletons. The jellyfish and worms that were so varied in the Ediacara fauna now had companions! (From *The Tower of Time*, a mural by John Gurche, courtesy of the Smithsonian Institution, Washington, D.C.)

CHAPTER 5:
SIFTERS IN A TROPICAL PARADISE

Australia's First Animals with Any Backbone

The Cambrian, Ordovician, and Silurian
570 to 408.5 million years ago

APPEARANCE OF ANIMALS WITH A BACKBONE

Alex took a break from a lecture being given at a scientific meeting in Adelaide and walked over to the South Australian Museum, just to have a look around their collections. There, in one of the dusty drawers, he found a small, red-stained piece of sandstone that had the imprint of a fish bone, a piece of armour plate. The label on this specimen read, "?Devonian, Mt Charlotte, Northern Territory, collected by D. J. Taylor, 1959". Alex thought to himself that the specimen looked like things he'd seen in Scotland - from fish that had no proper jaws - and that kind of fish had never been found in Australia before. After his discovery in 1968, Alex carried out a bit of detective work. He looked at a map of the geology of the area around Mt Charlotte in the Northern Territory and found that the rocks were not Devonian in age as the label had said, but, in fact, Ordovician, older than 439 million years. He was pretty excited about this. The fossil imprint could be from a fish that was one of the oldest backboned animals in the world, and most certainly the oldest one known from Australia. Alex chased up David Taylor, found out about other material that had been dug up at the same place and then eventually wrote a paper with Joyce Gilbert-Tomlinson telling the world about this weird and wonderful fish, Arandaspis prionotolepis, *our oldest backboned relative on the Australian continent.*

Figure 5.1. Studying specimens in museum collections can sometimes be a great way to discover a new species of animal or plant. The first fossil of Australia's oldest backboned animal, the 430 million-year-old *Arandaspis* was recognised in a museum drawer in Adelaide, long after it had been collected from a remote part of the Northern Territory. (F. Coffa)

Backboned animals (vertebrates), with a skeleton *inside* their bodies, rather than on the outside (as in grasshoppers) came late on the scene. Many other groups of animals were around before vertebrates appeared. We have some ideas about where they came from because some other animals seem to serve as links to animals *without* backbones (invertebrates), such as starfish, sea urchins and sea cucumbers. Next time you are wandering along the beach, remember that your origins lie

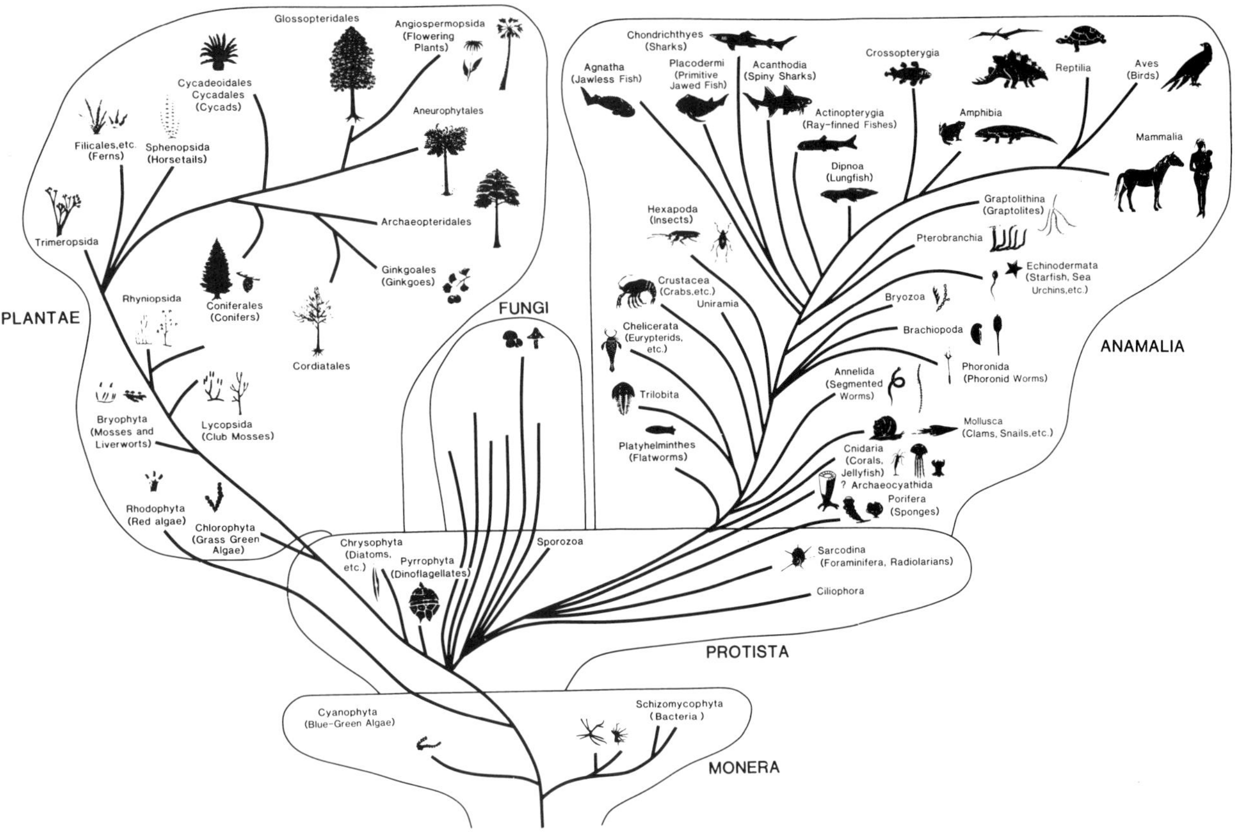

Figure 5.2. This family tree of living things shows how different plants and animals are related. The closest relatives of backboned animals, which include humans, are the starfish, sea urchins, sea cucumbers and graptolites - not snails and crabs or even the octopus.

ANCESTORS OF BACKBONED CREATURES

with the humble purple starfish you are staring at in the tide pool!

The animals that link us with the starfish are a mixed lot - *Amphioxus*, sea squirts, graptolites. *Amphioxus* is rather like a fish in shape. It has gills and sifts food from the water, perhaps not too unlike the ancient, jawless fish. Like us and our primitive backboned ancestors, it has a nerve chord that runs down its back, which is hollow and liquid-filled. *Amphioxus* doesn't have a "bony" backbone, but a stiff rod that acts like a backbone, called a notochord. The notochord runs the length of its back just under the nerve chord. Lots of primitive backboned animals had this too. Bits of this notochord can still be seen in animals we know today - it's the jelly-like stuff between the vertebrae in the backbone of a fish. In animals that have "bony" backbones, this stiffening rod is no longer important, so it is usually very small or lost completely.

Amphioxus uses its notochord rather like we use our backbone: to protect its nerve chord, to keep it from collapsing and as a place where it anchors its muscles.

With all these similarities to fish, it is quite surprising to discover that a youthful *Amphioxus*, its larva that is, looks like the larva of a starfish. *Amphioxus* has blood very similar to that

Figure 5.3. *Amphioxus*, the 'arrow-worm,' is a living invertebrate that is closely related to the backboned animals. This is the head end of a little animal that only grows to a few centimetres in length. The red lines are parts of the support structures for the gills that open to the outside world between the support bars. The supporting notochord lies above the gill support bars, and the hollow nerve network lies above the notochord. The notochord acts like our backbone and keeps the *Amphioxus* from collapsing into a heap. *Amphioxus* can swim about like a fish but spends most of its time buried in the sand with its head sticking out. There it draws water into its mouth and strains out food bits with its gills.

running through the veins of a starfish, or even a sea cucumber. In fact, if you could look down a microscope and watch a baby starfish, *Amphioxus* or cat grow from a single cell to something that begins to look a bit like you would see some very similar things. The tissue (called mesoderm) that develops into your gut or a starfish's gut develops in the same way, and this is quite different from the way it develops in a clam or grasshopper. All these similarities in development, construction and composition between starfish, *Amphioxus*, and the backboned animals (like the cat and us), strongly suggest that we are closely related. We are all closer "cousins" than any of us are to snails and worms!

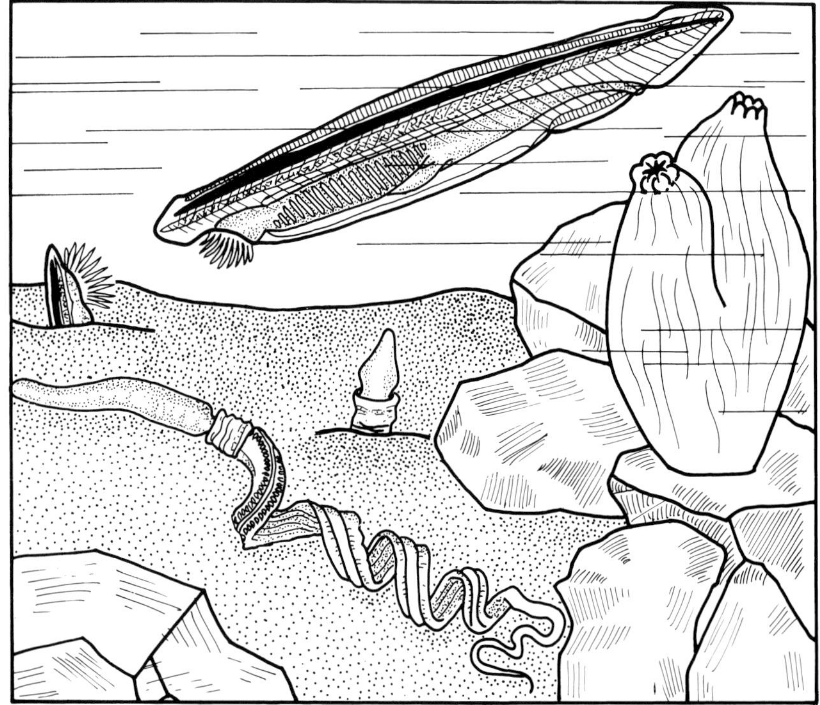

Figure 5.4. Three different animals without backbones that are close cousins of both the backboned vertebrates and the invertebrate starfish and its relatives. An *Amphioxus* swims overhead while another on the left is half buried with only its head end sticking out. An acorn worm lies on the seabed, and the head of another is sticking out from the sand in the middle of the picture. A sea squirt or tunicate is attached to a rock on the right. You are likely to come across the sac-like sea squirts on the beach at low tide or in tidal pools. When disturbed, they will squirt a stream of salt water at you, which, of course, is how they got their name. (D. Gelt)

Figure 5.5. Graptolites, whose name means 'writing on stone' are another group of 'spineless' animals that are probably related to humans. They are all extinct now, but there were many living in the seas when *Arandaspis* and other ancient vertebrates cruised about in the Ordovician and Silurian seas. Some were attached to the seabed or rocks; others floated about in the ocean's waters; all were sea-dwellers. Like today's colonial corals, graptolites lived in groups, and they were interconnected with a chord of living tissue, just as power lines connect our houses with a power station. Graptolites fed by sifting food out of the sea water, just like most of the other animals alive at this time. (I. Stewart)

Figure 5.5a. *Didymograptus*, an early Ordovician graptolite

Figure 5.5b. *Tetragraptus*, an early Ordovician graptolite

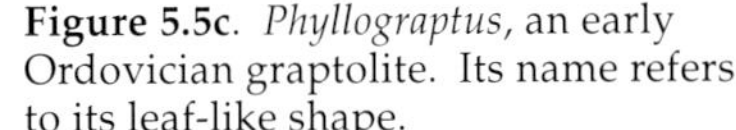

Figure 5.5c. *Phyllograptus*, an early Ordovician graptolite. Its name refers to its leaf-like shape.

Figure 5.5d. A diplograptid graptolite from the late Ordovician Period in Australia.

WHEN FISH WERE IN CHARGE

The *Arandaspis* that Alex Ritchie first found lurking in an Adelaide museum drawer was one of Australia's oldest backboned animals, and one of the world's oldest as well. Not only is it old, but it is one of the most complete very old backboned animals in the world, about 480 million years old.

Arandaspis wasn't large, only about 15 centimetres long. Its head and gill area was covered with a bony shield; it looked rather like an armoured tank. The rest of the body was covered by long, narrow scales, each decorated with a row of bumps. Unlike the fish we eat today, *Arandaspis* does not seem to have had any fins along the sides, top or bottom of its body. It just had a tail fin. It probably swam very clumsily, rather like you would if you put your hands by your side and kicked. It certainly wasn't very good at fast turns and screeching halts, but it was the best model around at the time, at least in the vertebrate world.

The only other Ordovician backboned animals known from Australia come from the same rocks as *Arandaspis*. They are also jawless, armoured fish. One is *Porphoraspis*. We don't know much about these fish, however, because our only clues are a few bits of bone preserved as imprints. We do know that their scales were different from those of *Arandaspis*. There are two other fish preserved in these ancient sea rocks, but such small fragments have been found that no one has yet given them a name. Another long field trip to Mt Charolotte might reveal many more of our most ancient backboned relatives!

Although Australia's first backboned animals are Ordovician in age, elsewhere in the world vertebrates had appeared in the Cambrian Period by about 500 million years ago. The jawless fish that were the first backboned animals were clumsy and sifted their food from the seas and streams like tea-strainers - there were no jaw-snapping carnivores or algae crunching herbivores - these came much later. Through the Cambrian, Ordovician and Silurian times "tea strainers" ruled in the shallow, tropical seas that bathed most of the continents of the world. These continents, including Australia, lay like beads on a string along the ancient equator. All seemed calm, warm and peaceful. It was the calm before the storm, however. Things were to change soon, rather dramatically.

Figure 5.6. (*Right*) *Arandaspis prionotolepis*, Australia's oldest vertebrate, was rather like a clumsy armoured tank, *without* the cannon. The head and shoulder armour protected it from some of the big predators of the time, the nautiloids, relatives of today's giant squid. (Courtesy of F. Knight and the Museum of Victoria)

Figure 5.7. This leopard tank has armour to protect it from the enemy, just as the armour of *Arandaspis* protected it over 400 million years ago.

F. KNIGHT 81

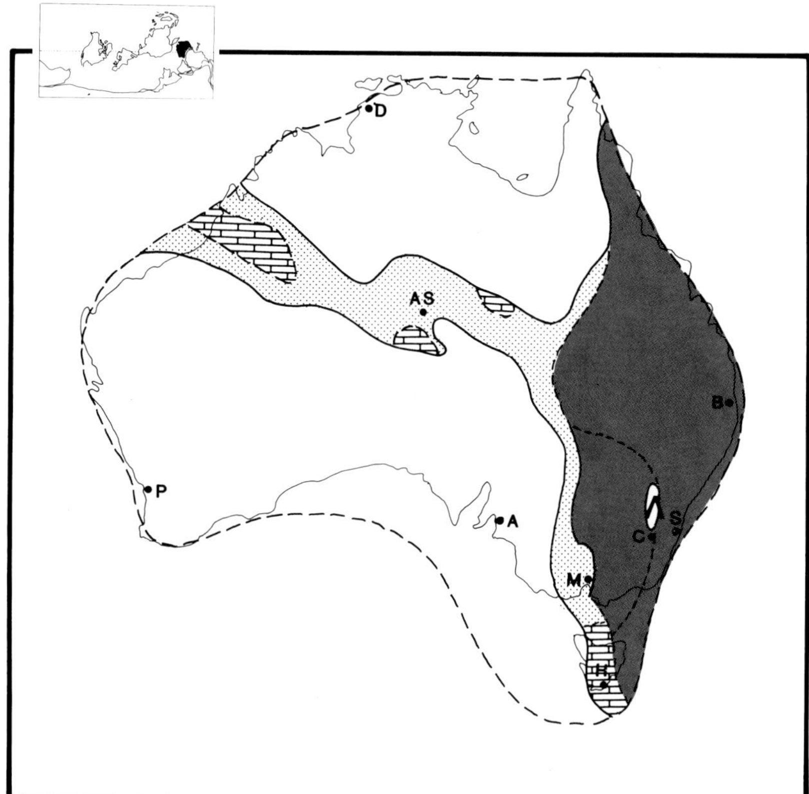

Figure 5.8. In Ordovician times Australia lay much farther north, on the equator and for much of the time its eastern segment was flooded by a deep sea. The shallow ocean that at times cut the continent in half was the home of Australia's first vertebrate animals. (D. Gelt)

THE ORDOVICIAN WORLD
510 to 439 million years ago

During the Ordovician, Australia's first backboned animals, the jawless fish such as *Arandaspis*, lived in the shallow sea that flooded onto the continent. At this time, all life was in the sea. Algae, including a variety of seaweeds, were quite varied.

Figs 5.9 and 5.10. A strange animal, *Notocarpos garratti,* related to starfish and sea lilies, from the Silurian marine rocks of Victoria. This specimen measures about 1.5 centimetres in diameter. Some palaeontologists think that this little invertebrate may be closely related to the first real vertebrates. We're not sure about this just yet - it's theory that needs more consideration and more fossils to prove or disprove its worth. (Courtesy of G. M. Philip)

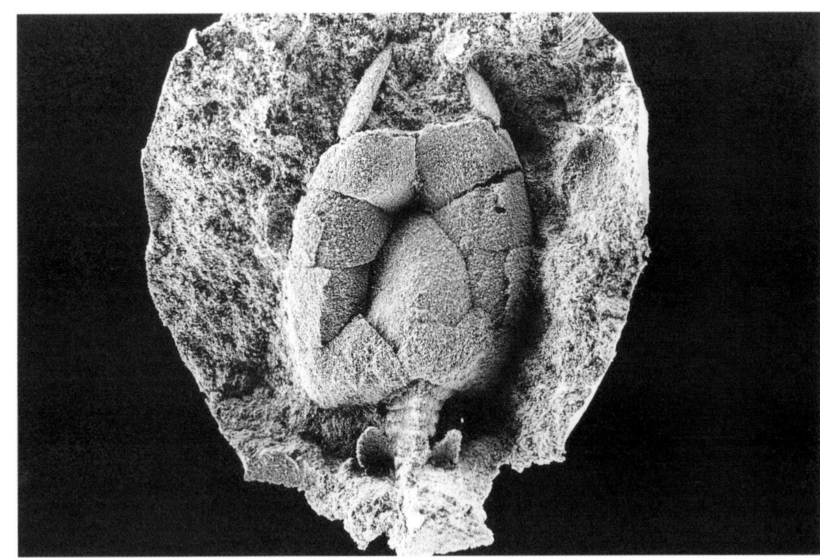

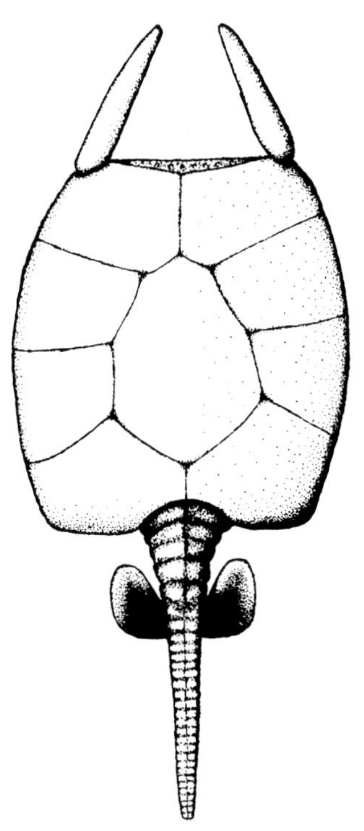

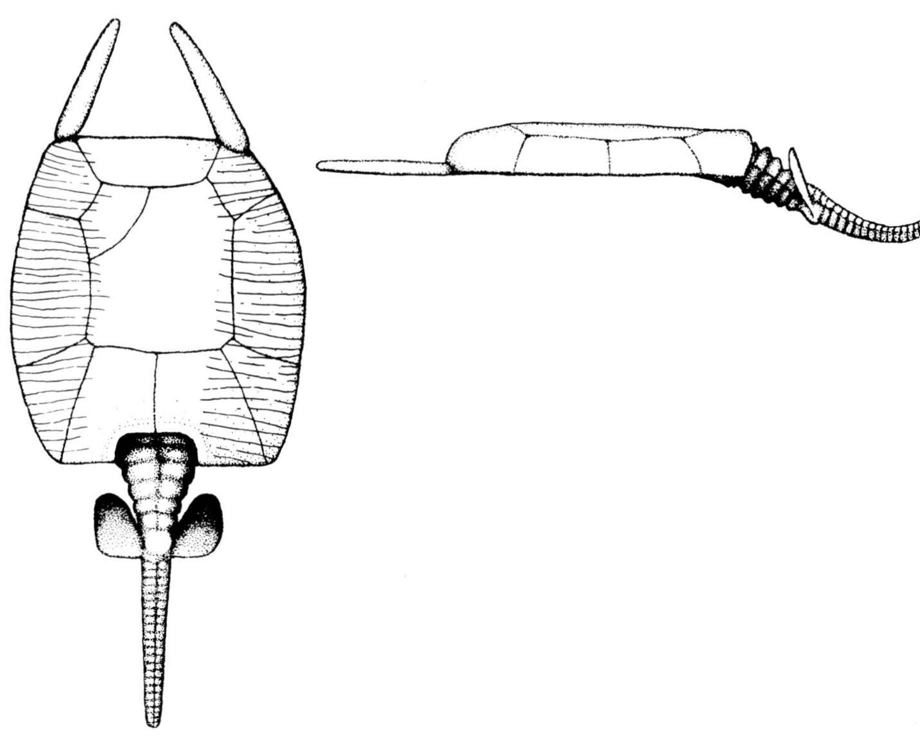

The Environment

Seas drained off the Australian continent early in the Ordovician, and mountains appeared in the south-east. Then, seas flooded in from the north-west and east, finally joining each other and splitting the continent into a northern half of low land and a rugged mountainous southern part attached to other southern continents. Sands were washed in from nearby land, and rising water in the seas deposited phosphates. The Stairway Sandstone, containing the first Australian backboned animals, was formed in this fashion. The sea in which the primitive vertebrates lived was full of new kinds of animals, especially those that sifted their food from the water, such as brachiopods and graptolites.

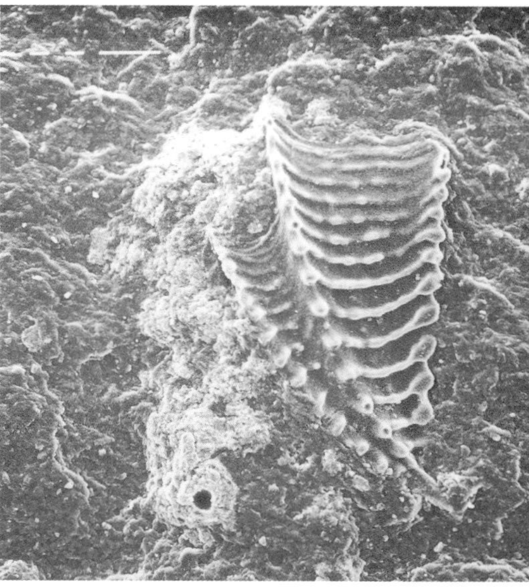

Figure 5.11. Other strange animals that roamed around in Ordovician oceans were the conodonts. We know of conodonts from tooth-like fossils such as this one, *Pygodos*, found in marine rocks in Australia. This tiny conodont is about 0.5 millimetres across! - and it is usually found in pairs. Scientists now think that the conodont animal didn't use these teeth to chew with but to support something like gills or a gizzard. We need to know a great deal more about these animals before we can draw a picture of them. (Courtesy of I. Stewart and J. Warren)

Figure 5.12. The early Palaeozoic seas (late Cambrian, Ordovician, Silurian) were rich in life. Trilobites burrowed, crawled and swam in these oceans. Nautiloids, the cone-shaped animals on the left, were among of the feared predators, as were the scorpion-like eurypterids, on the right and at the top. Corals and the clam-like brachiopods sifted their meals from the sea with their waving tendrils, and jawless fish joined these sifters with their tea-strainer feeding style.
(From the *Tower of Time* a mural by John Gurche, courtesy of the Smithsonian Institution, Washington, D.C.)

BEGINNINGS OF A MODERN WORLD – BIODIVERSITY ESTABLISHES ITSELF

The Plants

Algae were the only plants of Ordovician times. As far as we know they lived only in the seas and streams. These plants were of many kinds, ranging from one-celled plankton to the more complex kelp-like brown and green algae, and even encrusting red algae. Blue-green algae were still forming stromatolites, the massive calcareous structures so common in older rocks. The blue-green algae became more restricted during the last 500 million years of Earth's history because predators, such as grazing snails, destroyed what they built in all but the most severe of environments, such as extremely salty bays and lagoons.

The Backboned Animals

While trilobites and clams ploughed through the sea bottom, leaving traces from their burrowing and resting activities, called *Cruziana*, Australia's first backboned animals (such as *Arandaspis*) swam in the clear water above. The continental seas in which they lived were shallow, and had much oxygen - well suited for filter feeding organisms like these early jawless fish.

THE SILURIAN WORLD
439 to 408.5 million years ago

During Silurian times Australia was still attached to Gondwana. It moved slightly southwards, remaining close to the equator, however. Oxygen was now present in large enough amounts to form an ozone layer. This reduced harmful ultra-violet radiation reaching the Earth and favoured invasion of the land, first by plants. In Australia, the late Silurian *Baragwanathia*, and other primitive plants lived in swamps and marshes in many parts of the continent. No vertebrates of this age have been found in Australia yet, but there must have been fish in the seas, streams and lakes of the time.

Figure 5.13. A tabulate coral, also a common reef dweller in early Palaeozoic sea found near Buchan, Victoria.

The Environment

At the beginning of the Silurian, about two-thirds of Australia lay above sea level. The land in central Australia had been uplifted, as had parts of eastern Australia. In the east there were many water-filled ocean valleys, narrow coastal shelves, islands and erupting volcanoes. Late in the Silurian, shallow seas flooded onto the western part of the continent.

The Plants

The first true land plants in Australia appeared in the late Silurian. *Baragwanathia*, a club moss very similar in appearance to the living club-moss *Lycopodium*, was the most common plant. This Australian plant is one of the best preserved and earliest land plants from anywhere in the world. *Baragwanathia* and its relatives were small, all probably less than a metre in height. Many were low-growing herbs and others were lichens and related forms. All of these plants had evolved from algae and invaded an environment that was very unpredictable and dry, not always wet like the one they had come from. To cope with the dry conditions, these plants developed special woody tubes (xylem) through which water moved from the ground to their stems and leaves. These tubes also gave support, helping plants to stand erect and present their leaves, spines and stems to the sun and allowing them to produce their own food. They reproduced by spores, not seeds, for seeds had not yet developed.

In Australia, Silurian floras grew in wetlands beside the sea; some *Baragwanathia* plants are preserved in marine rocks, and plants were evidently being washed out of the swamps into the sea, where they were deposited.

The Backboned Animals

At present no vertebrates are known from Silurian rocks in Australia. This is most probably because the fossils have not yet been found. Elsewhere in the world, a variety of fish lived in the seas, freshwater streams and lakes. It is quite likely that they also did so in Australia. *One day* someone will find their bones here.

CHAPTER 6:
REVOLUTION OF THE JAWS

Fish, Fish and More Fish.
Attack on the Tropical Seas and Streams

The Devonian
408.5 to 362.5 million years ago.

The rain was pouring down in sheets as it often does in Victoria during May. It was wet and miserable and the rocks, a dark black, were slippery and even blacker than usual. The sky was thick with cloud, and it was growing darker all the time. John was swinging his hammer gingerly at some mudstone rocks that stuck out along the edge of Freestone Creek, near Blue Pool. His hammer blows could be heard for a kilometre or more and it had attracted a local farmer. "Lookin fer gold are ya, mate?" he asked John. "Nah, fossils - you know, the remains of old animals" John replied. "I think there's fish in here." The farmer gave John a rather puzzled look, thinking - "fish in those rocks? This guy must be crazy." Still, he seemed a nice enough bloke and so John was invited up to the farmer's house for a cuppa when he had finished his rock bashing.

Figure 6.1. The Devonian, 408 to 362 million years ago, was the 'Age of Fishes', a time when fish developed true jaws. Jaws allowed fish to eat a great variety of foods they couldn't eat before, so jaws led to an explosion of new forms - hundreds, even thousands of new species, (from *The Tower of Time*, a mural by John Gurche, courtesy of the Smithsonian Institution, Washington, D.C.)

John Long was really pleased with such an invitation and after another hour or so digging up a wonderful specimen of a fish, he turned up at the farmer's house for some tea and a drying-out in front of an open fire. What he had to show the farmer was something entirely new to him and to science, a new species of fish that had never been seen by anyone before. This fish was a placoderm; it had a thick coat of bony armour surrounding its head and shoulders, but more importantly, it had *jaws*, very different from the head of *Arandaspis*, the fish Alex Ritchie had studied from the Ordovician of Mt Charlotte in the MacDonnell Ranges.

Jaws were the great invention of the Late Silurian and Early Devonian periods. They allowed their new owners, the fishes, to try out a lot more kinds of foods than they had been able to before. Before having jaws, fish had been "tea strainers" or "vacuum cleaners," able only to sift particles and small bits and pieces out of the water. Now, with jaws, they could graze on plants, attack other fish and slice pieces off them, or kill them and then eat bits, and they could scavenge for things on the sea or stream bottoms. With jaws came the development of many

Figure 6.2. An armour plate of a placoderm fish from the Devonian of Victoria. The decoration on these armour plates can be so unusual that individual species can be recognised from a single fragment. (J. Long)

different kinds of fins - ones along the back and belly, on the sides. Tail fins were redesigned so that they increased lift - the fish could move upwards as well as straight forward or down in the water. Fish were really in a new design stage and new models were coming off the production line at a great rate in the Devonian. By the end of the Devonian, all the major fish groups, including those that gave rise to our modern fish, had developed - from lampreys, sharks, and lungfish, to the modern ray-fins. It was the hey-day of fishes - and nothing much except fish was around. The world was theirs on a platter (instead of them being on the platter for the world)!

John showed the farmer his new fish and almost convinced him that this fish's Victorian home and the farmer's land had been under water some 380 to 400 million years ago. It had been a place of fast moving mountain streams flowing down off the slopes of erupting or dormant volcanoes. The fish John had found had a tall crest running down the middle of its head and shoulder armour, looking a bit like a sail. John later decided that the "sail" might have served as a stabiliser to keep the fish upright as it swam in those fast flowing streams. Whatever its purpose, the crest or sail made this fish *Bothriolepis cullodenensis*, the Crested Box Fish, very different from any of its relatives, and so John eventually gave it this new name in a scientific paper.

The Devonian Period has been called "The Age of Fishes," because it had so many different kinds of fish - not only in Australia but the world around. In Australia there are a few places that are very rich in fossil fish bones. At Taemas-Wee

Figure 6.3. Mountain streams and lakes in Victoria were alive with fish during the Devonian. Very common were the antiarch placoderms, especially *Bothriolepis*, an early jawed fish with heavy armouring. It had many, small pointed teeth on its jaws. It probably fed by waiting on the bottom of a stream or pond and lunging at anything that came along with a quick flip of the tail. (Drawing by F. Knight, courtesy of the Museum of Victoria, from Rich, van Tets and Knight, 1985)

Jasper near Canberra, in south-eastern Victoria, in central New South Wales, and on Gogo Station in the Kimberley region of Western Australia, fossil fish are abundant and often very beautiful.

Perhaps the most fantastic place to fish for fossil fish is on Gogo Station. It's a fairly flat area today with *Acacia* bushes scattered around, a place you might not be too impressed with at first, that is until you realise just what you are walking across. That low ridge that pops up about a half kilometre away is the edge of an ancient reef. In fact, you are walking across the sea bottom of what was once an ancient lagoon, more than 360 million years old. The rocky ridge is made up of the remains of corals and algae and other marine animal fossils that once protected this lagoon from all but the stormiest of waters. Fish lived and died in the lagoon and their bones settled on the bottom. Some time after their death, carbonate-rich waters deposited lime around their skeletons and they were enclosed in rocky nodules. It took the sharp eye of a museum director and his determination, together with that of an ace collector from the British Museum, to realise what a gold (or should we say "goldfish") mine Gogo was. David Ride and Harry Toombs discovered that fish skeletons were in those nodules and that nodules were lying all over the ground at Gogo: paradise, at least for a fossil fish catcher, had been found! At first, getting the fossils out of the hard nodules seemed nearly impossible, but acetic acid (the acid in vinegar) was the answer. By soaking the nodules in five parts of acetic acid to 95 parts of water, the carbonate rock around the bones was slowly eaten away. Complete, beautiful and very delicate skeletons were prepared, and they could be studied, almost as you would study the skeleton of one of today's fish after the platter and the fish have been cleaned.

Figure 6.4. Tropical seas and massive reefs like this one in the South Pacific today were present in western Australia in Devonian times. Even further south near Buchan, Victoria, small reef-like structures are known to have existed.

Figure 6.5. A Devonian placoderm from the Gogo locality in western Australia. This was an arthrodire, a kind of placoderm that had a joint between its head and shoulder armour. It had no teeth but instead had armour plates on the outside of the jaws and these acted like knife blades. These fish were very frightening predators, some growing to several metres long. Their armour would have protected them from other, perhaps larger fish, who were also hunters. (J. Long)

AUSTRALIA ON THE EQUATOR AND AN AGE OF FISHES

The Gogo reef was home for a lot of fish. The armoured placoderms were there. *Bothriolepis*, the armoured fish that looked a bit like the living armoured catfish in pet shops today, lived there too. Big placoderms, called arthrodires because they had a hinge between their head armour and their neck armour, were the predators in these waters. *Bothriolepis* was mainly a bottom feeder and took whatever came along. Then there was a variety of lungfish, some with duck bills, which made a meal by pulverising water plants, small water-dwelling animals and a certain amount of mud and grit that got scooped up in the scramble for food. The fan-like teeth of these fish acted like choppers, with several blades slicing past one another, up and down.

Figure 6.6. Fishes of the late Devonian Gogo Reef in Western Australia. This reef built along the edges of the Kimberley land mass about 360-370 million years ago. An arthrodire, *Harrytoombsia* (left centre, named after the British Museum fossil collector who first collected at Gogo Station), is hot on the fins of several smaller, primitive ray-finned fish called *Moythomasia*. Another arthrodire, *Holonema* (lower left), slowly swims towards a long-snouted lungfish, *Griphognathus* (lower right). Two other arthrodires can be seen above, the snub-nosed *Ctenurella* (top) and the unicorn-snouted *Rolfosteus* (middle), showing how many different kinds arthrodires existed at the time. *Bothriolepis*, an antiarch of the placoderm group more often found in fresh waters, swims between *Ctenurella* and *Rolfosteus*. (Drawing by F. Knight, courtesy of the Museum of Victoria, from Rich, van Tets and Knight, 1985)

On these same Gogo reefs were the beginnings of the fish that are so numerous today, the ray-finned forms whose fins, as their name implies, are supported by many bony rays that attach to a bony fin base on the body of the fish. These were to become the most adaptable and varied of all fish groups. They were alive already in the Devonian, and thousands of species in this group exist today - such as the barramundi, sailfish and trout of our tables.

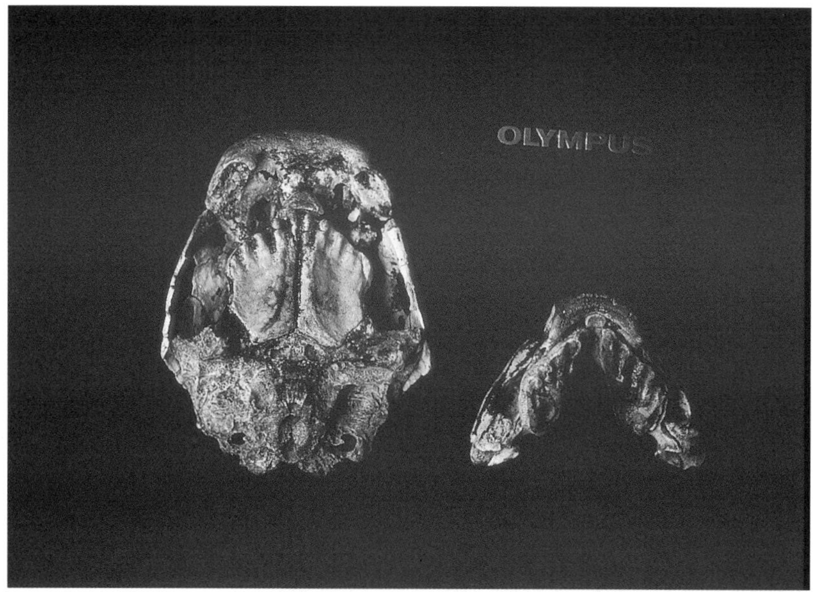

Figure 6.7. One of the Gogo lungfish, a snub-nosed *Chirodipterus australis*. The fan-shaped teeth of the upper jaw (right) slice into the fans on the lower jaw, acting like several knives chopping at the same time. The teeth of lungfish have remained much the same for more than 350 million years. The teeth of the modern Australian lungfish of tropical Queensland are not all that different from those of this Devonian fish. These teeth are excellent for slicing up water plants into a digestible mush. (J. Long)

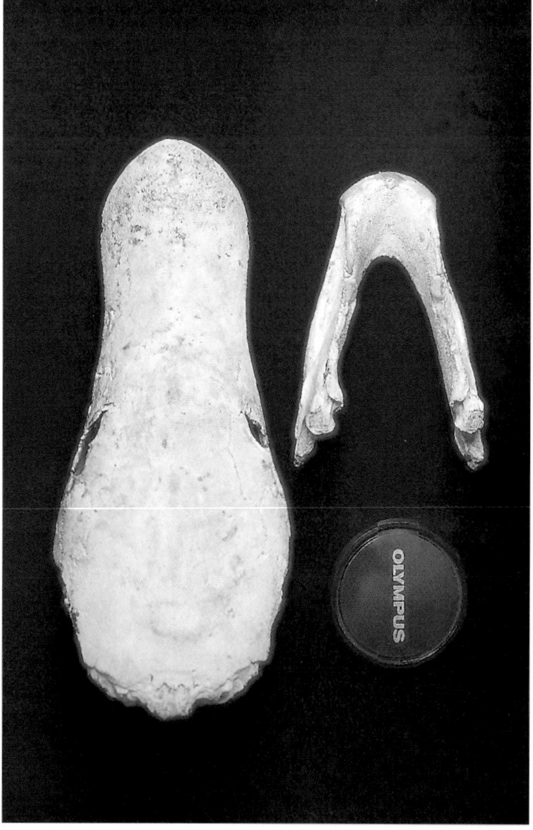

Figure 6.8. Another Gogo lungfish, *Griphognathus whitei*, has a skull that is very like that of a duck with its broad, short bill. The feeding habits of this fish may not have been all that different from those of some ducks today. This is an example of the shape of bones adjusting to the kind of food being eaten, no matter what an animal's ancestors were like (called convergent evolution). (J. Long)

Not all groups of fish that existed in the Devonian have left living relatives. The placoderms and acanthodians (or spiny sharks) are extinct now. Only a few species of jawless fish (the lamprey and hagfish) and lungfishes still survive. The Age of Fishes came to an end, but fish are still the most numerous vertebrates in our world today, even though a single group of fishes dominates. These ray-finned fish began in the Devonian and never looked back. Sharks were not quite so successful, but are also still around today, and they too had their beginnings in Devonian times.

We wonder if the farmer to whom John showed his fossils realised that those ancient fish were the great, great ...... great grandparents of those that swim in his Freestone Creek today. Sometimes we don't even realise that what we're walking on today was something totally different in ancient times, but had those fish not existed in the past, we might not even exist to be here digging up their bones!

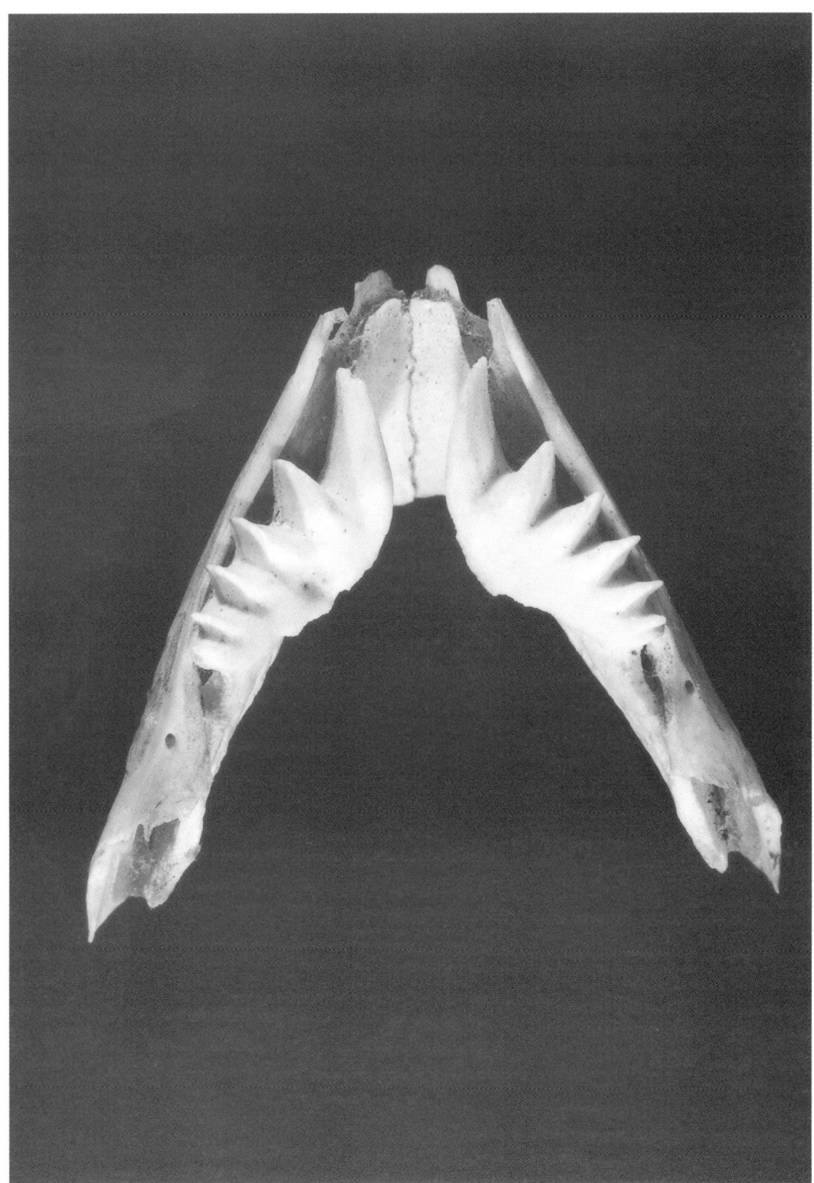

Figure 6.9. Lower jaw of a lungfish, showing how the fan-like teeth fit in. The slicing action is an up and down motion, and the jaw moves very little back and forth. (A. Kemp)

Figure 6.10. An acanthodian, or spiny shark, from the Devonian of Australia. This fish called *Culmacanthus* was found at Freestone Creek in Victoria. The spiny sharks are well-named because they have very large spines in front of many of their fins. Acanthodians were not really sharks but a separate group of fish that started in the Silurian and were extinct by the early Carboniferous. Some people who are studying fossil fish think that the acanthodians may be the ancestors to the ray-finned fishes, but we have much more to learn before we can be sure. (J. Long)

THE DEVONIAN WORLD
408.5 to 362.5 million years ago

By Devonian times most of Australia was above sea level. There were some exceptions, such as the reefs that developed in the Kimberleys in north-western Australia. There were other reefs along the east coast, where volcanic activity and earth movement created mountains. Devonian plants lived in swamps and along lake, river and ocean edges, and in places where the water table was high. Vertebrates included marine and freshwater fishes and amphibians, the first vertebrates to invade the land.

Figure 6.11. Australia and the world during the Devonian. Seas only touched the Australian continent along the east coast and at points along the west. Big river systems flowed across part of Queensland, and New South Wales and Victoria. Much of Australia was above sea level, and the northern parts still lay close to the equator. Australia was still a part of Gondwana: it was not a separate continent.

OCEAN AREAS
- Shallow seas
- Shallow seas with limestone deposits
- Shallow seas with reefs (light blue)
- Shallow seas with glaciers
- Deep ocean

LAND AREAS
- Dry land
- River deposits
- Coal swamps
- Glacial deposits
- Volcanic deposits
- Volcanoes

The Environment

Still a part of Gondwana, Australia lay in the tropical regions through the Devonian. Compared with previous times, the continent was fairly dry, as widespread salt deposits show. In general, however, temperature and humidity were about the same as they are today at the same latitudes. In the early Devonian, earth movements and volcanic eruptions occurred along the east coast, and deep oceans often separated volcanic islands there. Later in the Devonian, mountains were formed, and seas drained off the continent. In the very late Devonian, big reefs built of stromatophoroids (calcareous colonial organisms), algae and corals grew in north-western Australia, and these have preserved a great variety of organisms, including

Figure 6.12. Receptaculitids were sponge-like animals that lived on the reefs around Buchan, Victoria, during the Devonian. Their nearest relatives are still not known. (S. Morton)

vertebrates. Curiously, the coral faunas were rich in individuals but had few species, suggesting that the area may have been close to the limits of reef growth. In areas near the equator, where reef growth is best, lots of species exist but fewer individuals per species. Today that area is about 20 to 30 degrees of latitude north or south.

Figure 6.13. Brachiopods were 'tea-strainers' - they sifted plankton from the tropical seas to make their meals. This group looked a bit like clams, but differed from clams because the two shells in each pair were not mirror images of one another. Many different brachiopods were living during the Palaeozoic. (S. Morton)

The Plants

The club moss *Baragwanathia* lived on into the early Devonian in south-eastern Australia. By the mid-Devonian, tree-forming club mosses, like *Lepidosigillaria*, formed impressive swamp forests. By the late Devonian another giant club moss, *Leptophloem*, was abundant, and forests were still growing in and near fresh waters. These forests may have been made up of plants that lived much the same way as mangroves do today, with their root systems in the swamps.

The Backboned Animals

Most Devonian vertebrates were fish, and during this period they developed into a great variety of types. The big reefs of the north-west served as home to the first jawed fishes, the placoderms, such as *Bothriolepis*. But, the spiny acanthodians and ancient lungfish were also there. Rare jawless fish were represented by the thelodonts, the most recent occurrence of this group anywhere in the world - everywhere else they had gone extinct.

The brisk mountain streams of Australia's south-east also contained a variety of fish, mainly armour-plated placoderms such as *Groenlandaspis* and *Bothriolepis*. These fish were very similar to those in southern China, which may have lain very close to Australia at this time.

An important happening at this time was the first appearance of amphibians, who left their footprints at two places in south-eastern Australia. Those from the early to mid-Devonian rocks of the Grampians in Victoria may be the oldest record of amphibians in the world. A single jaw of a labyrinthodont, *Metaxygnathus*, from the late Devonian of Australia, suggests that identification of the tracks is correct. The second set of tracks was found in the eastern part of Victoria, set in late Devonian sands.

Figure 6.14. A single scale of an ancient fish photographed by an electron microscope. This scale is about 1 millimetre across. One fish would have had several thousand of these scales. (S. Turner)

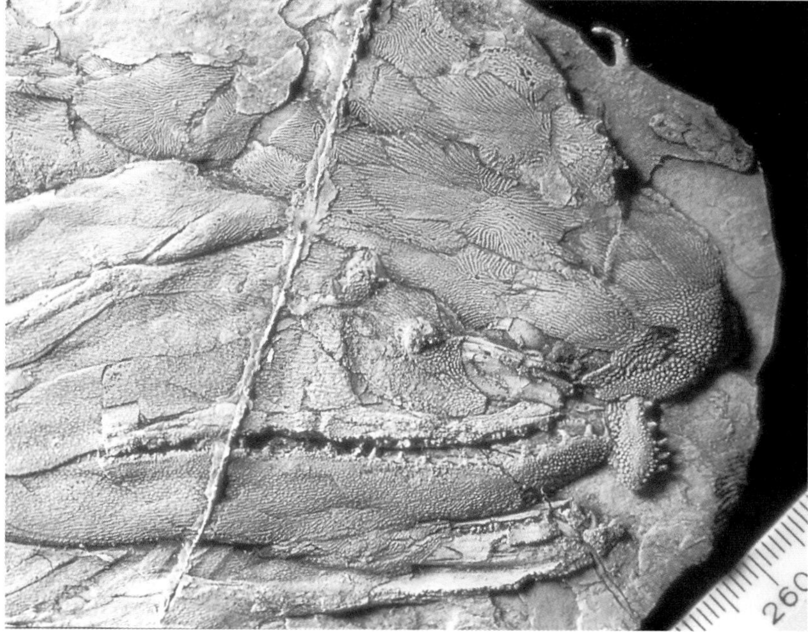

Figure 6.15. A latex rubber impression taken of a fossil fish from the Devonian of Victoria. Some fossils are easier to study if a cast or mould of the fossil is made. This fish is a palaeoniscoid, one of the oldest ray-finned fish known. Fish like this were the ancestors of our modern fish, from trout to sailfish. The palaeoniscoids were active, fast-swimming fish with big eyes and a mouthful of sharp teeth. They were small to medium-sized predators of the day.

CHAPTER 7:
LAND AHOY!

Plants and Amphibians "Slither" onto Land

The Devonian
408.5 to 362.5 million years ago

Imagine what the land would be like if there were no plants, no trees, no flowers, not even a little patch of moss. You've just imagined the entire earth, above sea level, from about 420 million years ago back to when the world began, some 4500 to 4600 million years ago. It must have been a rather dreary place, a bit like the surface of Mars now, only the temperature on average was a bit warmer than on Mars and the atmosphere was different. It would have been a place without much colour, without green especially, and not very alluring to insects and backboned animals. Soils would not have been forming as they are today. It would have been a very different world from the one you step out into each morning.

Figure 7.1. *Siderops*, a labyrinthodont amphibian. (Drawing by F. Knight, courtesy of the Museum of Victoria)

The world, on land, really began to change remarkably in the Silurian, and some of the first changes may have taken place right here in Australia, in Victoria. Some of the first plants to live successfully on land were probably algae, which may have inhabited moist soils, but there was a limit to what such plants could do. They couldn't grow tall, as they really had nothing to give them support, like the support a skeleton gives. So algae, even when associated with fungi, like in lichens, stays quite small. No algae forests have ever existed, unless, of course, the algae stayed in the water, and giant seaweed "forests" developed with a little help from the buoyancy of the water.

So, if you were a plant wanting to make a career out of living on land, you had to solve a few of these problems. In order for plants to invade land they had to avoid drying out, and so cuticle developed as a thickened outer skin to prevent water loss. Successful land plants also had to have some sort of support to keep from collapsing into a flat blob. They also had to transport water and some nutrients from the soil to the rest of the plant. These problems were solved by the development of woody vessels called xylem (and pronounced "zilem") that both transported water and gave support to the stems of these aspiring young *"terrestrialians"* - that is *"land-lovers"*. Somewhat later, specialised cells developed that produced continuous layers of xylem. This allowed the girth or thickness of the stem to increase and a new kind of transport tubes- phloem - to develop, bringing nutrients from the soil to the rest of the plant.

PLANTS INVADE THE LAND

Size became no problem after this, and forests grew from Devonian times onwards.

The only other major hurdle to jump, if you were a plant heading for the good life on land, was reproducing yourself in such an unpredictable, unfriendly environment. Reproduction had always taken place in the water or in moist terrestrial environments before. Now, all that had to change. Spores were the answer, spores like ferns have today. The small brown spots on the back of the leaves are spore concentrations, and when the wind blows some of these minute spores off the plant, thousands of them are carried off by the breeze. They are produced on one plant, and they in turn give rise to tiny plants that produce eggs and sperm. These still have to live in moist environments to reproduce, but the spore-bearing plants that result from the union of the sex cells of the male and female plants can survive under drier conditions and, from the root systems alone, can actually increase their size and even regenerate after a fire or drought. So, even though these early plants were not completely able to survive under really dry conditions, they were on their way.

Keeping moist, supporting yourself and being able to produce more like yourself were the big issues in surviving on land. Several plants managed to take care of all these issues in Silurian and early Devonian times. The first were the psilophytes. Today*Psilotum* is a rare survivor of this group; it lacks leaves or even specialised roots. It keeps life simple. Yet it could stand up straight and colonise the land enough to add a splash of green to the late Silurian landscape. Soon after, came a number of other spore-bearing plants, some of which have a few surviving relatives. Among these early groups were the lycopsids or club mosses, such as *Baragwanathia*, many species of which had a special diamond-shaped pattern on the stems, scars where branches had once grown and dropped off. Another group was the sphenopsids, or horsetails, whose leaves are arranged in little skirts or whorls surrounding a jointed stem, and attached at each joint. Ferns appeared early, too, and they are still abundant today.

All of these plants had one thing in common. They produced spores. Seeds had not yet made their appearance, and until they did, plants were tied to moist environments during at least one part of their life.

By mid-Devonian times the land, at least around the edges of streams, lakes and the ocean, were greened by plants. The stage was set for the entrance of the amphibians, and for that matter, insects and land snails as well. Plants provided a source of food, protection from the heat, as well as protection from something that might be waiting to eat these new land-dwellers. Plants provided a new place that had not been used by animals before, and so new jobs were available and new niches were open - at least for a while. For the animal that could move into such an environment, there were many advantages. Imagine the surprised look on the face of "Joe Arthrodire" as he was chasing you if you could just hop out of the pond! He couldn't follow! This may have been one of the

Figure 7.2. *Psilotum*, a living relative of the first land plants that are known from the Silurian. The psilophytes had no leaves and were very simple plants. (B. Furher)

first real pressures that "encouraged" amphibians to develop - from the fish that gave rise to them - that such a move would be a good one. The food that was "there for the taking" might have been a lure too.

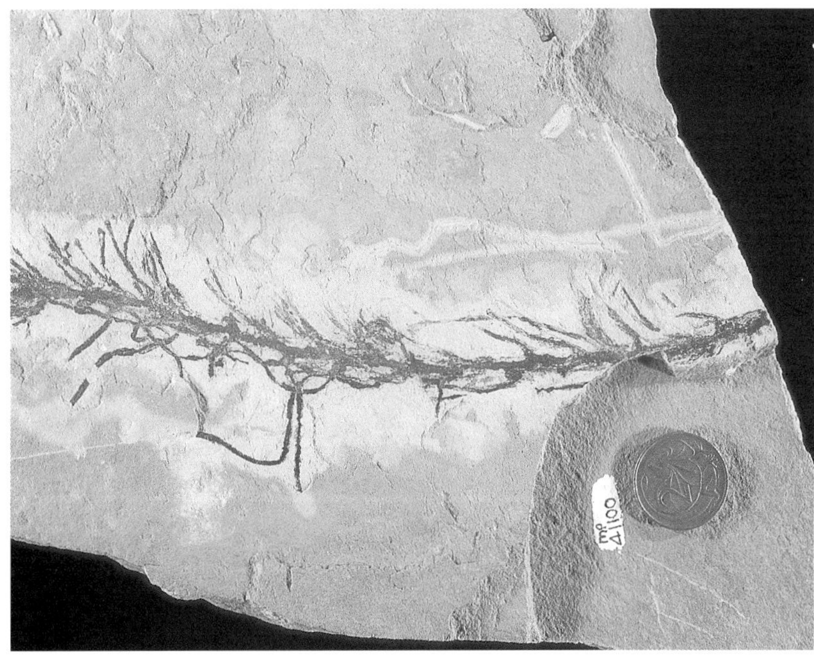

Figure 7.3. *Baragwanathia*, an early land plant from Victoria, from late Silurian and early Devonian times. This was one of the club mosses, a lycopsid, a group now represented by small herbs only. Silurian lycopods were not very large either, some only as high as a metre or so, but they grew into enough swamp vegetation to tempt amphibians onto the land. (S. Morton)

Figure 7.4. *Selaginella*, one of the few surviving club mosses today. These small, ground hugging plants are all that remain of the tree forming club mosses of the Devonian and Carboniferous, a time when spore-bearing plants ruled, before the seed-bearing ones took command.

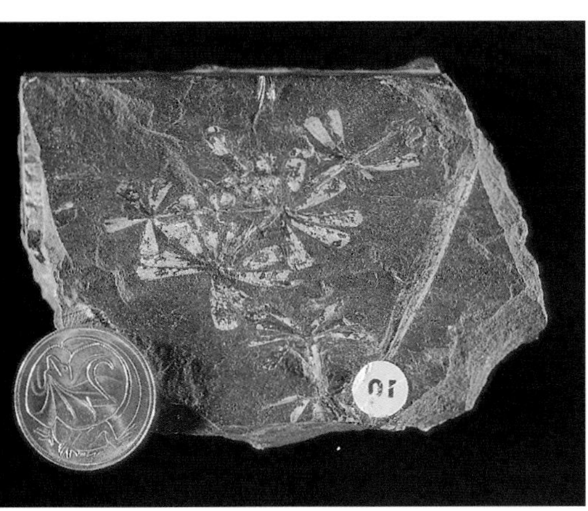

Figure 7.5. This fossil sphenopsid is a relative of the present-day*Equisetum* a horsetail rush that rarely grows over half a metre tall. In Devonian and Carboniferous times, gigantic forms closely related to this sphenopsid formed the forest trees. They too bore spores and not seeds. The whorls or 'skirts' of leaves can be seen on this specimen. (S. Morton)

One group of bony fish, the crossopterygians, seems the best bet for a group that might have given rise to the first land dwellers, the labyrinthodont amphibians. Both the crossopterygians and the labyrinthodonts had teeth with complexly folded enamel, and both had skulls that looked a lot alike, with a joint in their braincase (only in the most ancient of the labyrinthodonts). Just thinking about such a braincase hinge gives us a headache. Crossopterygians also had fins that contained the basic building-blocks of a foot. It's not too difficult to imagine how you could turn a crossopterygian fin into a foot with a few minor architectural modifications.

Crossopterygians even had a little sac called an air bladder, which was originally used to help them stay in the right part of the stream, much as you might use a rubber tyre to keep yourself where you want to be in the water. This air bladder stored air. When crossopterygians moved onto the land, it is likely that this same air bladder eventually evolved into the lungs needed to deal with oxygen in the air rather than oxygen in the water.

Figure 7.6. Skull of *Siderops*, a labyrinthodont amphibian belonging to the first group of vertebrates to venture onto the land. The top view of the skull shows the kind of pattern that is so distinctive for labyrinthodonts - many lines on heavy bones. Their skulls tended to be fairly flat, and their mouths were full of teeth. Labyrinthodonts were probably predators and spent time both in and out of the water. (A. Warren)

Figure 7.7. Skeleton of a labyrinthodont amphibian. The head was usually large and had grooves, which may have been the location of tissue sensitive to water pressure. Fish have such grooves. In our world today, only animals that live in water have such a canal system, so the conclusion has been that labyrinthodonts spent some of their time in water, or a very close relative of theirs did! (A. Warren)

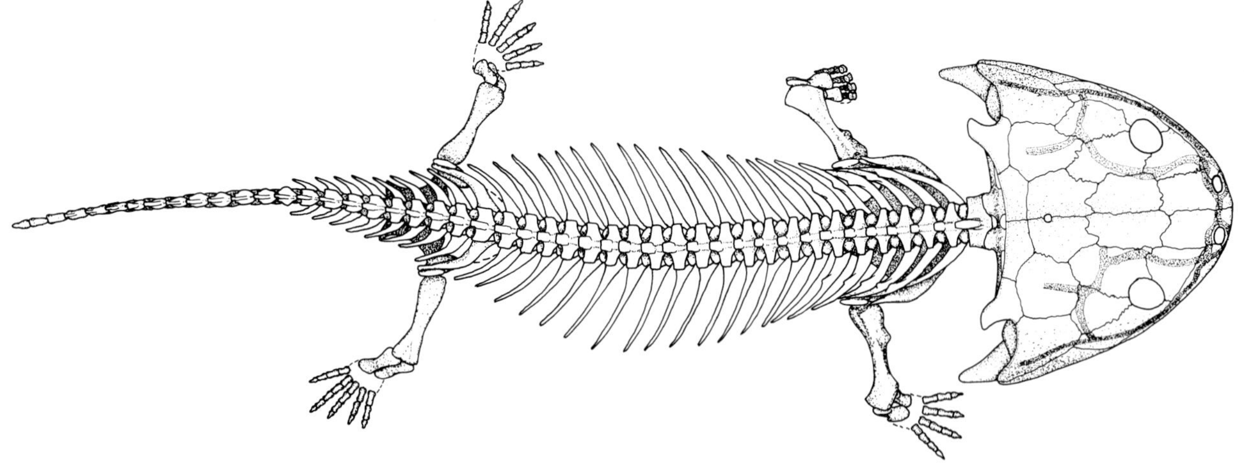

Amphibians had to face the same problems as plants in order to take on a land dwelling existence. They developed a thicker, more protective skin to avoid drying out, they coped with breathing oxygen from the air rather than water, they developed tongues and other accessories that allowed them to take in water; and they developed ways of reproducing in a land environment. They still laid their rather fragile eggs in moist environments; however, like the spore-bearing plants, they had not completely freed themselves from the water.

Australia's oldest amphibians, and perhaps the oldest in the world, come from the Grampians in western Victoria. The fossil remains of these first amphibians are a set of footprints left by a small land-dwelling vertebrate during the early to middle Devonian. The slab of rock with those footprints was found as one of the paving stones covering a spacious front yard. Late Devonian amphibian tracks are also known from the Genoa River area, and a jaw of a labyrinthodont, called *Metaxyganthus* is known from the late Devonian of New South Wales, near Forbes, where the big radio telescope is located. So, by the end of the Devonian, vertebrates were quite certainly off and running, or perhaps slithering much of the time!

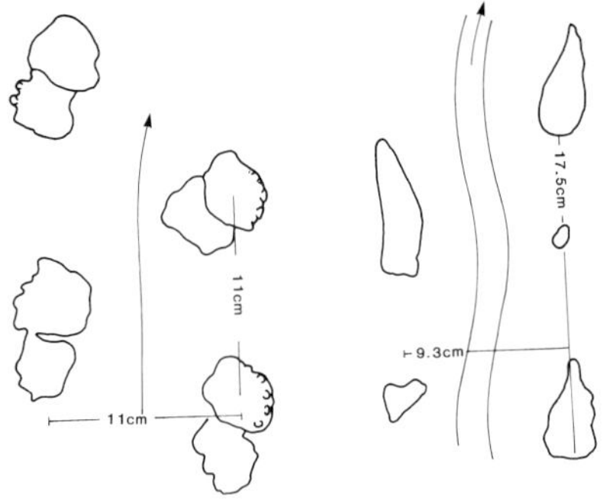

Figure 7.8. Two sets of tracks thought to have been made by land living vertebrates, amphibians, from the late Devonian of Genoa River, eastern Victoria. The tracks on the right show that the animal was dragging its tail, whereas the amphibian on the left was not. (J. Warren)

Figure 7.9. Tracks of an amphibian from the late Devonian of Genoa River, eastern Victoria. (S. Morton)

CHAPTER 8:

REFRIGERATION DE LUXE, AN ANTARCTIC ADDRESS

Australia Cruises to the South Pole. The Late, Great Palaeozoic Ice Age, and then Someone Turns on the Heater

The Carboniferous, Permian, and Triassic
362.5 to 208 million years ago.

Lea hopped out of the car and looked out across the green countryside near where her parents had stopped for lunch. Spring was finally here after a dreary, wet Melbourne winter, and it was great to see the sun and to watch the fields turn from yellow brown to a new green. The car was parked at what looked like a roadside rest area, but there were no signs to announce that it was a rest place. It looked like a rest area because the ground was so smooth and level. The rocks looked as though a big bulldozer had passed over them. Lea bent down to run her fingers across the smooth surface. It looked almost as though it had been polished, like the stones she had seen worn down in Iris's rock tumbler - a little machine used to make stones smooth for jewellery. The rocks here beside the road not far from Bacchus Marsh were smooth and polished, she thought to herself. Then she noticed something queer. The polished rocks had many tiny grooves and scratches on them, even though most of their surface was quite smooth. Many of these grooves ran parallel to each other, like the wire of a fence or train tracks. What in the world had caused them? Maybe someone had carved them there; maybe Aborigines or someone else. But how could they carve so many? And why? She tried carving a bit in the rocks with her Swiss army knife, but found it difficult to make any scratches. Bits of the knife steel were left on the rock instead - so the rock was harder than the knife! "Whoever did this carving must have been pretty strong!"

Had her parents not been geologists, Lea might have gone on wondering for a long time about such polished rocks with their grooves and scratches, but she soon had an interesting explanation. During early Permian times, around Bacchus Marsh in Victoria and in several other parts of Australia as well, there

Figure 8.1. Life moved on to the land in the Carboniferous. Amphibians appeared in the Devonian and were followed by reptiles in the Carboniferous. As far as we know, Australia had no reptiles until the Triassic, but the continent's poor fossil record may be responsible for this impression. Club moss forests are what greeted these early backboned creatures. This was all to change towards the end of the Palaeozoic Era when cold temperatures wiped out these primitive forests. (From *The Tower of Time*, a mural by John Gurche, courtesy of the Smithsonian Institution, Washington, D.C.)

were huge glaciers hundreds of metres thick. As the climate cooled and the ice became thicker, glaciers slowly moved across the ground, smoothing off and polishing the rock surfaces beneath them. These huge glaciers acted like garbage trucks and bulldozers at the same time. As they moved across the countryside they picked up debris, and rocks became stuck on their underside. So, as the glaciers slowly ploughed across the land surfaces, always down towards the sea, the accompanying rocks scored the grooves in the Earth's surface beneath, the weight of the glacier providing the force that scored.

Glaciation began in Australia sometime during the late Carboniferous and continued into the early Permian. Besides this being evident from the kinds of rock that were laid down at that time, and the polished and grooved rock surfaces, the change to cold conditions is also clearly shown by the kinds of plants that were abundant. The club moss forests of warmer times (the Devonian and early Carboniferous) gave way to low, rather widely spaced-seed fern heaths, similar to heaths you might see today in the polar tundra areas of northern Europe or southern Argentina, or even a bit like the vegetation in cold coastal areas now. Times were tough, and there were only a few species of plants that could live under these conditions, so there were very few species even of the hearty seed ferns during the late Carboniferous.

When the glaciers were largest, the only backboned animals that left fossils behind in Australia were fish, as far as we know. Amphibians, *must have* been lurking somewhere in the ice-free areas, but we have yet to discover where. Only after the cold turned to cool, and as widespread coal swamps appeared in places like Sydney and Newcastle, did amphibians begin leaving their skeletons to be fossilised. *Bothrioceps*, a labyrinthodont amphibian, for example, is known from the Newcastle coals.

What caused all this cold and the massive changes in the forests - in fact, a complete loss of the forests for awhile? The answer is quite simple - Australia had moved. It moved from its position near the equator in the Devonian almost to the South Pole by the late Carboniferous, a rather incredible journey, covering tens of degrees of latitude! In addition to this rather impressive "holiday" that Australia and other parts of Gondwana took, world climate was cooling (we are not quite sure of the reasons). All in all, Australia was probably not a very hospitable place in late Carboniferous and early Permian times. This was to change - later.

By late Permian and early Triassic times, however, conditions were a great deal warmer. Amphibians thought this was great. In fact, during Triassic times Australia had a wider variety of labyrinthodont amphibians than any other part of the world. It was a labyrinthodont's heaven and *haven*, as we will see later. In the early and middle Triassic seven of the 10 known families of these sorts of amphibians lived right across Australia - from the north-west to the south near Hobart, and from Sydney to south-eastern Queensland. They came in many shapes and sizes - rounded headed, long-snouted similar to crocodiles, and many other shapes in between. These animals

Figure 8.2. Scratches on the surface of a pebble once caught beneath a moving glacier. (S. Morton)

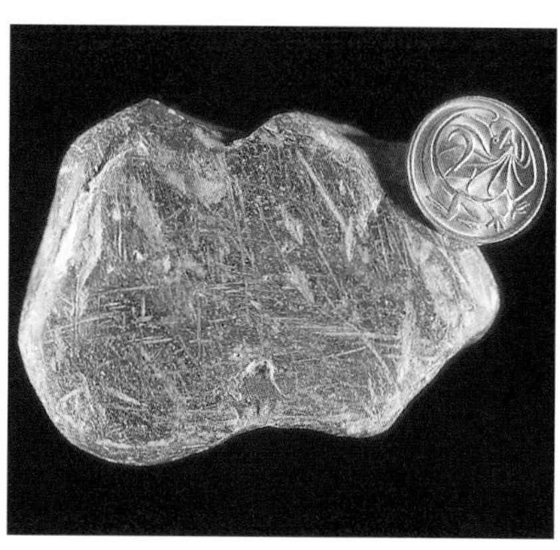

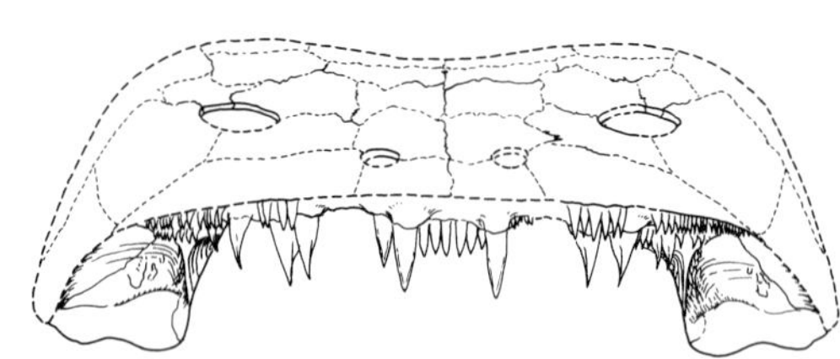

Figure 8.3. This is what it was like to stare at a labyrinthodont amphibian in the mouth. This collection of many sharp teeth of many different sizes was an awesome prospect, really terrifying for a smaller vertebrate. (A. Warren)

COLD TEMPERATURES LOW SEA LEVELS AND MAJOR EXTINCTIONS

were most likely the small, medium and large carnivores of the day both in fresh waters and on the land around the swamps. Their hundreds of sharp teeth would have been a most depressing sight for a small animal!

Labyrinthodonts were the only amphibians of their day. They lived on in Australia for a long time after they had become extinct elsewhere in the world - into the Jurassic and even Cretaceous times. Australia was a real haven for this group, perhaps because it was cut off from much of the rest of the world in some way, even in Jurassic times. Not until much later did the only group of amphibians that live in Australia today - the frogs - reach this continent. Their first bones left here are only about 20-25 million years old.

Fish are known to have lived in Australia from the Carboniferous to Triassic periods even through the lowest temperatures of glacial times. The spiny sharks, the lobe-finned crossopterygians, likely ancestors of the amphibians, and primitive ray-finned fish were here. So, too, were sharks. One interesting shark, *Pleuracanthus*, lived in fresh water rather than in the sea where most sharks live now and lived in the past. Australia was a haven for these backboned beasts, too, and they lived on into Triassic times, even though they had become extinct everywhere else by Permian. When alive in Australia in the Triassic, they were "living fossils!" Everywhere else they were extinct and preserved only *as* fossils. Lungfish, with their very strange fan-shaped teeth, left many of their fossils in the freshwater sediments of the Triassic, as did ray-finned fish, which were not as primitive as they had been in the Permian. They began to look more like the fish you might catch today.

Reptiles were living in Australia in the Triassic. Their bones have been found alongside those of labyrinthodonts in places like south-eastern Queensland, but such finds have been rare. Perhaps this is because reptiles didn't spend much time near the water (very different from the amphibians) and so had less chance of being preserved. Reptiles had cut their ties with the water. They had a tough, scaled skin and laid eggs quite capable of surviving in dry environments, just like the eggs that chickens lay now. These reptiles were the *first* backboned animals able to spend their entire lives out of water, their birthplace. So, the way was paved for dinosaurs.

THE CARBONIFEROUS WORLD
362.5 to 290 million years ago

Early in Carboniferous times, the northern part of Australia was still near the Equator. The climate was fine for the giant club moss forests that lived in extensive swamps. Everything changed in the late Carboniferous. Australia steamed south rapidly, the climate cooled, and seed ferns took over in the glaciated regions.

The Environment

Australia embarked on an incredible journey in the Carboniferous, moving from the equator to near the South Pole, and saw a

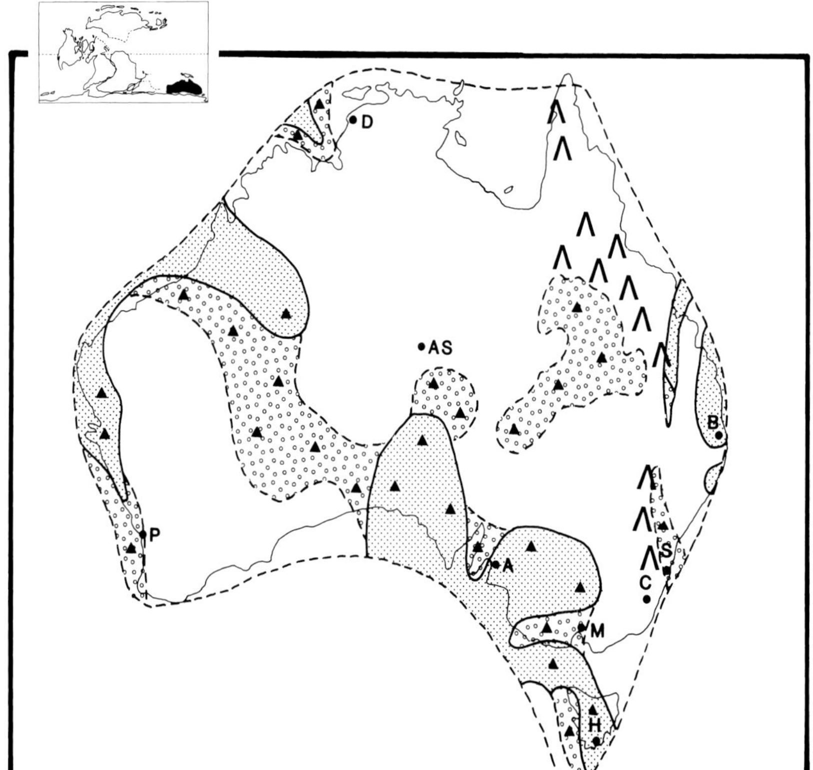

Figure 8.4. What Australia and the world looked like in the late Carboniferous.

OCEAN AREAS

- Shallow seas
- Shallow seas with limestone deposits
- Shallow seas with reefs (light blue)
- Shallow seas with glaciers
- Deep ocean

LAND AREAS

- Dry land
- River deposits
- Coal swamps
- Glacial deposits
- Volcanic deposits
- Volcanoes

change from warm to icy conditions. At the start of the period, shallow seas flooded areas of both the east and the west, depositing their limey sediments; volcanoes were blowing up all along the east coast. At the beginning of the late Carboniferous, Australia moved very quickly southward, and temperatures became quite cold. First, mountain glaciers formed in the volcanic mountains of the east, and then big continental glaciers developed, eventually covering large parts of Australia.

The Plants

On land, the plants clearly show that major climatic changes had occurred. The giant club moss *Lepidodendropsis* was common in the early Carboniferous in the swamp forests, where horsetails grew like rushes on water margins together with the first plant to produce seeds, a small-leafed seed fern. The seed ferns seemed to thrive as the glaciers advanced, bringing with them very cold temperatures which wiped out the club moss forests. Only a few plants survived, such as *Rhacopteris*, a seed fern with a very primitive leaf structure. The few species that existed, grew close to the ground, tundra-like. At times, when the climate was warmer, club mosses (*Cyclostigma* and *Subsigillaria*) and other seed ferns (*Cardiopteris*) moved back to Australia.

The Backboned Animals

The Carboniferous vertebrates known from Australia are mainly fish. The redbeds of the Mansfield Basin in south-eastern Australia deposited in fresh water lakes and streams, for example, contain spiny sharks (acanthodians), lungfish and lobe-finned fish (crossopterygians) that may be ancestors of amphibians, as well as *Elonichthys*, a palaeoniscid related to our modern ray-finned fish. Primitive shark fossils also come from marine rocks in the north-west. Even though all fossils of vertebrates from the Carboniferous of Australia are those of water-dwelling forms, animals must have been living in the forests on land. Unfortunately, the fossil record here is not complete enough to tell us what was going on in Australia at this time.

THE PERMIAN WORLD
290 to 245 million years ago

Australia lay close to the South Pole at the beginning of Permian times and was covered by glaciers. Climate warmed up during the Permian; ice caps melted, and sea levels rose. *Glossopteris* (a seed fern) swamp forests were all over the place in these cool times, and these gave us many of our great coal deposits in eastern Australia and near Perth. A variety of fish lived in lime-rich seas and in rivers that flowed into these coal swamps; labyrinthodont amphibians sneaked around the east coast coal swamps in search of a fish dinner.

The Environment

In the early Permian, Australia was still in the grip of a major glaciation. It was extremely cold! Separate glaciers occurred in the south-western, central, south-eastern and north-western

Figure 8.5. What Australia and the world looked like in the Permian.

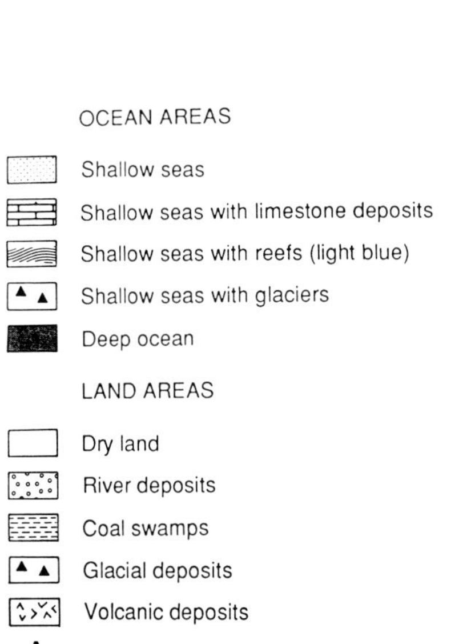

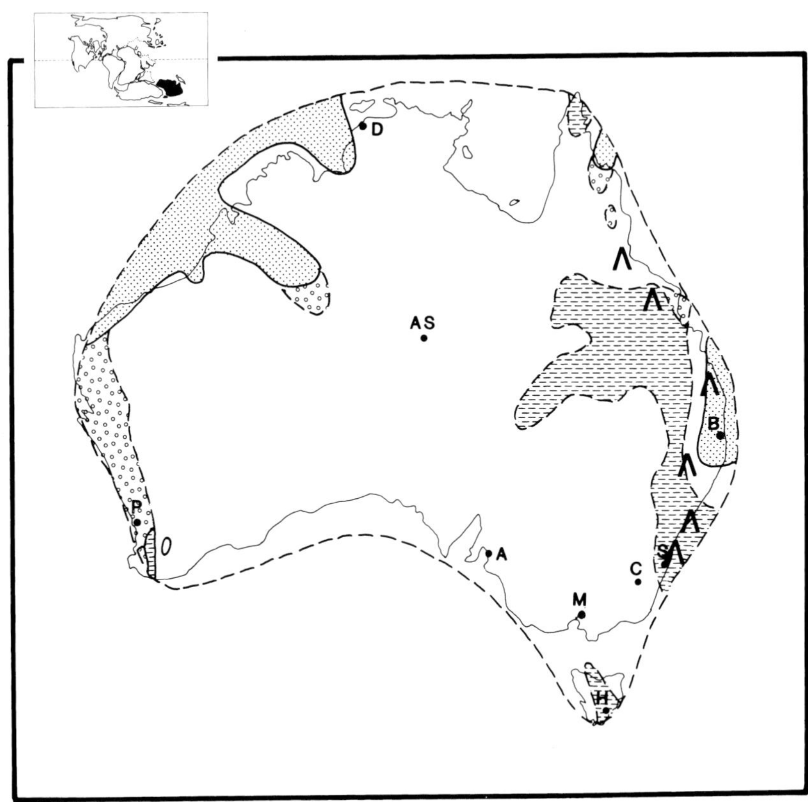

parts of Australia. Evidence for the ice is found in the deep parallel grooves ground by the boulders trapped at the base of glaciers, as well as in the big piles of rock debris (morainal debris) dumped by glaciers as they began to melt.

Sometime in the early Permian, the ice began to melt. Ice caps shrank, and the sea level rose, so that the sea invaded the area around Sydney and parts of eastern Queensland, as well as parts of the south-east coast. As sea levels dropped in the late Permian, many of these areas became swamps and bogs in which major coal deposits formed, rather than lagoons.

The climate changed from polar to cool, and temperatures rose, although it remained fairly wet through all of the Permian.

The Plants

On land the *Rhacopteris* tundra plants stood aside for new forms as the climate warmed. Seed ferns were still varied, and many of these lived right across the big Gondwana landmass, indicating that the Australian and parts of the Asian continents must have been connected at the time. *Gangamopteris*, with its oval leaves, was one of the first new, warmth-loving plants to appear, probably growing in the milder coastal areas. *Gangamopteris* forests may have occupied environments similar to those of the birch-rich taiga of the northern hemisphere polar regions today.

This "palaeotaiga" gradually gave way to the *Glossopteris*-dominated swamp forests, after the permafrost (a permanently frozen part of the soil) was no longer present. Permafrost would have destroyed the roots of the seed fern *Glossopteris*, which contained large open chambers. *Glossopteris* and ginkgoes were deciduous, that is they dropped their leaves each year, but the conifers alive at this time, cordaites, were evergreen. Australian conifers of this period show growth rings, evidence that the climate was seasonal. The smaller club mosses, horsetails and ferns formed the undergrowth in these Permian forests, and after burial and compaction they gave rise to many of our black coal deposits.

The Backboned Animals

Despite lots of plant fossils being preserved in Permian rocks, only one vertebrate, a labyrinthodont amphibian called *Bothrioceps major*, is known in rocks of this age. Quite often, conditions favourable to preserving plants are not good for preserving bone - conditions are too acidic. So far, the good record of amphibians and reptiles found in other parts of Gondwana for this time has not been repeated in Australia. It is probably only a matter of time before such fossils are found, however.

Both marine and freshwater fish are known from the Permian, including a variety of sharks. Freshwater fishes lived in the swamps that produced the Newcastle coal deposits; primitive ray-finned fishes with especially big eyes were the most common.

THE WORLD HEATS UP, DRIES OUT, AND REPTILES RULE

THE TRIASSIC WORLD
245 to 208 million years ago

Climates warmed a bit compared with conditions in the late Permian, but stayed cool and wet into the Triassic. Most of Australia lay above sea level, and lake and river deposits of this age are widespread. Seed ferns, ginkgoes, and conifers were the main plants, while many different labyrinthodont amphibians, and both lizard-like and mammal-like reptiles, lived everywhere from tidal flats to broad river plains.

The Environment

Australia stayed near the South Pole, but in a humid climatic belt. Rivers were flowing over much of north-western Australia and south-eastern Queensland (the Rewan Group rocks). The Hawkesbury Sandstone of Sydney was laid down on a broad floodplain, as was the Knocklofty Sandstone in southern Tasmania near Hobart. Large deltas, swamps, rivers and lakes were familiar sights. However, coal swamps, although existing, were not abundant, in contrast to Permian times. Fossil soils of the early Triassic in Sydney were formed under cool to temperate conditions, perhaps not so seasonal as in the preceding Permian.

The Plants

Plants changed at the beginning of the Triassic. At the end of the Permian, the climate warmed and dried, the coal swamps dried up and the *Glossopteris* seed ferns disappeared. Replacing

Figure 8.6. What Australia and the world looked like in the Triassic.

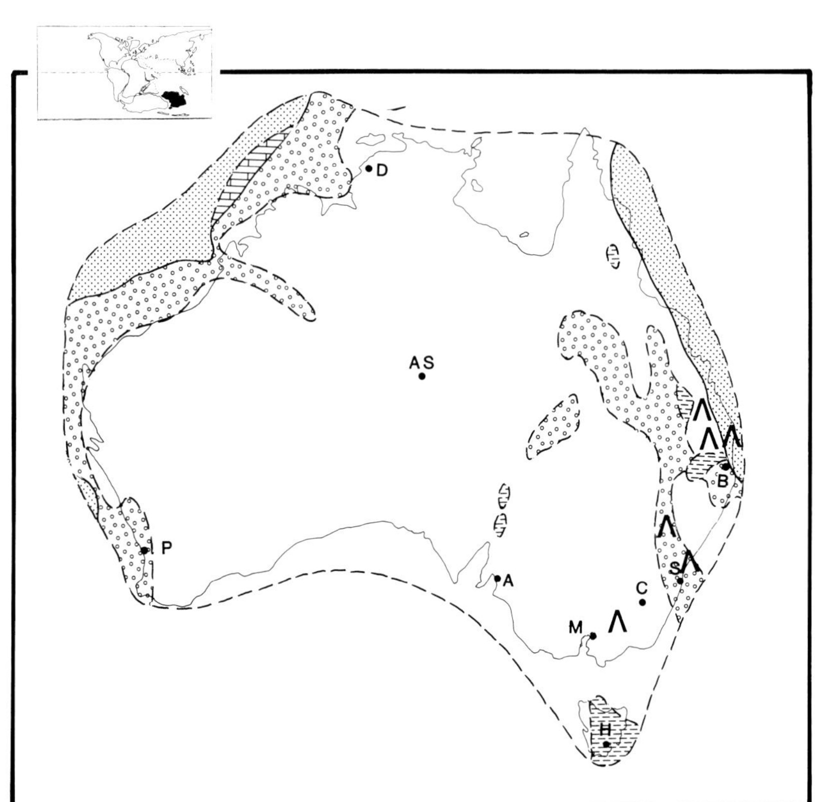

these was another seed fern *"Thinnfeldia" callipteroides*, which lived on river flats. Conifers with small leaves, horsetails and club mosses with whorls of strap-like leaves were the new revolutionaries that took over. Later still the forests were formed of conifers and *Dicroidium*, another seed fern - both belonging to a Gondwana-wide flora that lived in many different kinds of habitat. Ginkgoes were common towards the end of the Triassic, and the podocarpaceous conifers made their first appearance.

Backboned Animals

The freshwater lagoons and swamps of the Triassic were ideal homes for many species of labyrinthodonts and reptiles, mainly known from the redbeds of the Rewan Group rocks (especially the Arcadia Formation) of Queensland, the Knocklofty Sandstone of Tasmania, and the Blina Shale of Western Australia. Nowhere else in the world were labyrinthodont amphibians as varied as they were in the early Triassic of Queensland.

Rare mammal-like reptiles, dicynodonts, also lived in south-east Queensland in the Triassic, as did many different kinds of small, lizard-like reptiles, including *Kadimakara*. The first of the dinosaurs, a large carnivorous theropod, left tracks in Queensland in what is now called the Blackstone Formation.

Fish lived in the streams and marine waters of the Triassic: palaeoniscoids, more modern looking ray-fins and lungfish, including *Ceratodus*.

Figure 8.7. (*Right above*) The life story of brachiopods is mainly a Palaeozoic one. Some still survive, but the greatest variety lived from Cambrian to Permian. Because of what happened in late Permian times, mainly because of worldwide glaciation and cooling - it wasn't a good time for brachiopods.

Figure 8.8. (*Right centre*) Chonetid brachiopods that lived in ancient Permian seas of Western Australia. (N. Archbold)

Figure 8.9. (*Right below*) Spiriferid brachiopods from rocks deposited in Permian seas in Western Australia. Cold temperatures and low sea levels during much of the very late Palaeozoic brought about the extinction of many groups of sea-dwelling animals that were dependent on the shallow continental shelf seas for a home. With the sea level down as much as 300 metres, not much of the continental shelf was under water. Brachiopods were a group severely hurt by these conditions. (N. Archbold)

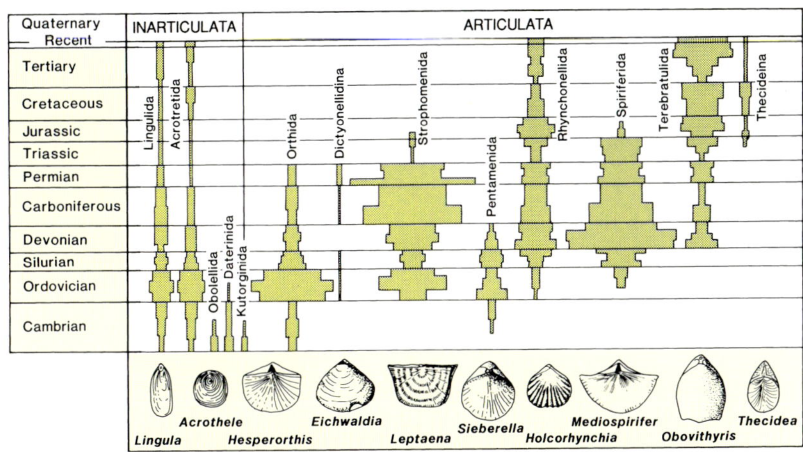
Quaternary
Recent
Tertiary
Cretaceous
Jurassic
Triassic
Permian
Carboniferous
Devonian
Silurian
Ordovician
Cambrian
INARTICULATA
ARTICULATA
Lingulida
Acrotretida
Obolellida
Daterinida
Kutorginida
Orthida
Dictyonellidina
Strophomenida
Pentamenida
Rhynchonellida
Spiriferida
Terebratulida
Thecideina
Lingula
Acrothele
Hesperorthis
Eichwaldia
Leptaena
Sieberella
Holcorhynchia
Mediospirifer
Obovithyris
Thecidea

CHAPTER 9:
ENTER THE DINOSAURS

Australia's Successful Failures: Polar Dinosaurs, Cool Sea Monsters, and Big Troubles 65 Million Years Ago

The Jurassic and Cretaceous 208-65 million years ago

Rebecca attached the final wire to the complex web of red cord that entwined the cliff face and disappeared into the dark shaft which cut into the sea cliff just west of Cape Otway, on the south coast of Victoria. She turned and walked away from the cliff, blew a small whistle, letting us all know that the cove (Dinosaur Cove) was to be cleared for blasting conditions. A thorough look across the expanse of the shore platform and cliff face proved to her that everyone had followed the often repeated routine, which had produced the two, 8-metre-deep shafts into the cliff in just over a month. All was truly clear. Only then did Rebecca attach the wire to the "joy box", a relatively innocent looking black cube that had a small crank and a button. Count-down began —10, 9, 8, 7, 6, 5.... and then the crank was turned furiously to build up the charge...4, 3, 2, 1, and the button was pushed! A deafening roar washed across our small group just milliseconds after a plume of smoke and flying rock shot seaward at an amazing speed out of the eastern shaft. More rock flew out of the shaft in one blast than we had been able to dig up with hand tools in nearly two weeks last season, when we had no explosives.

Many fishermen have asked Rebecca what she was digging for - gold? oil? The answer is always the same – dinosaurs. Rebecca, the mine manager, has searched for minerals all over the world, including Thailand, where when the local 'warlord' stopped her buying dynamite she used sugar and fertilizer as explosive. On holiday and a volunteer mine manager for the Dinosaur Cove Expedition, she always seems to gain some small pleasure from the generally surprised reaction of the questioning passerby.

Figure 9.1. The late Palaeozoic and the Mesozoic, a world dominated by the reptiles. (from *The Tower of Time*, a mural by John Gurche, courtesy of the Smithsonian Institution, Washington, D.C.).

VICTORIA'S POLAR DINOSAURS

The operations at Dinosaur Cove are a bit drastic for getting at dinosaurs; they are certainly less drastic, however, than the destruction wrought by the sometimes massive seas that crash

Figure 9.2. Eric, a black-ticket blaster, setting up for blasting at Dinosaur Cove on the south coast of Victoria. (F. Coffa)

Figure 9.3. Gelignite is used to blast away the rocks acting as a dinosaur tomb at Dinosaur Cove, west of Cape Otway, Victoria. (F. Coffa)

Figure 9.4. At isolated Dinosaur Cove bones of early Cretaceous dinosaurs, about 106 million years old, have been found in ancient stream channels and lakes. (F. Coffa)

Figure 9.5. Sending equipment into Dinosaur Cove on a flying fox. (F. Stewart)

DRIFTING CONTINENTS AND POLAR DINOSAURS

onto these southern Australian shores. In fact, it is the force of the sea that has made southern Victoria one of the very best dinosaur hunting grounds in Australia. The sea's continuing erosion of the coastline has exposed wide areas of nearly flat lying rocks - sandstones, siltstones, mudstones, even conglomerates - the remains of a giant stream complex that flowed out across a broad, steep-sided valley 105 to 120 million years ago. This was at a time when Australia and Anatarctica were beginning to pull apart like hot cheese, but before the Great Southern Ocean spilled into the lowland left between them. That would happen a few million years in the future, a time the dinosaurs would never know.

But why are the methods needed to collect the bones of these dinosaurs so drastic? The reason is the hardness of the rocks. Once the sandstones that now serve as the tomb of dinosaurs were soft sand; the mudstones, soft mud and so on. But after the dinosaurs died, the sand and silts containing their bones sank just as the valley itself sank with the further separation of Antarctica and Australia. The same is occurring in East Africa today as the eastern part of Africa is attempting to rip off from the rest of the continent. The East African Rift Valley has developed now just as the southern Australian Rift Valley began 120 million years ago. Further north in Africa today, the Red Sea is an example of a sea invading a rift valley. This is what happened in Australia during the very last part of most dinosaur history, in the late Cretaceous and early Tertiary.

Figure 9.6. Getting people into Dinosaur Cove using a rope. The flying fox is too dangerous. (F. Coffa)

Figure 9.7. Hand tools are best for breaking up smaller pieces of rock at Dinosaur Cove. (F. Coffa)

Figure 9.8. Small pieces of rock must be looked at carefully to be sure that no important bones or teeth are missed. (F. Coffa)

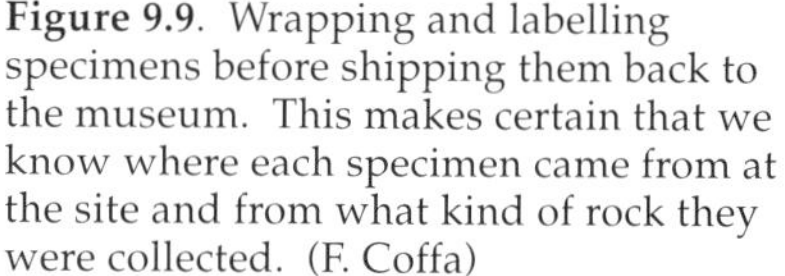

Figure 9.9. Wrapping and labelling specimens before shipping them back to the museum. This makes certain that we know where each specimen came from at the site and from what kind of rock they were collected. (F. Coffa)

Figure 9.10. Work at Dinosaur Cove is done using air-powered mining tools provided by Atlas Copco, a Swedish mining equipment company. The wooden portal over the mine entrance is to protect diggers from falling rocks. (F. Stewart)

Figure 9.11. Excavating at Dinosaur Cove with a Panther Rock Drill. (F. Coffa)

Figure 9.12. Drilling holes at Dinosaur Cove for putting in rock breaking wedges. It's not very clean work! (F. Coffa)

Figure 9.13. The dinosaur jaw shown in Figure 9.17 after it had been cleaned up. (F. Coffa)

Figure 9.14. This is what southern Victoria may have looked like in Cretaceous times — a fern-laden, rift valley which was cool, wet and totally dark three months of the year.

As the sands and silts sank, they in turn were covered with more sand and silts, which continued to increase in thickness. As these sediments built up, their weight compressed and heated the sediments underneath them, some as deep as 1.5 km below the surface. Hot liquids containing certain minerals rich in carbonates (chemicals that make up part of bicarbonate of soda and sea shells) and silica (which makes glass) flowed through some of these sands and silts and eventually they became hard rock. In other words they became *lithified*. Unlike some of the more famous dinosaur localities in the world, such as those in Colorado or Wyoming, or even at the Flaming Cliffs in the People's Republic of Mongolia, Victorian localities are in hard rock, and you simply must set up a limited mining operation in order to recover them. This is just what has been done at Dinosaur Cove in cooperation with the National Parks and Wildlife Service, the Victorian Department of Labour, and hundreds of interested volunteers from Australia and many other parts of the world.

Many people are quite surprised to discover that Australia has dinosaurs of its own. The big sauropods from the American West or the spur-pawed *Iguanodon* from England and Europe are generally more familiar forms. But Australia does have its own dinosaurs, and over the past 20 or so years finds of new Australian dinosaurs have been increasing. Amateur collectors play an essential part in many of these discoveries and will continue to to do so for a long time in the future. It is quite possible that you might find your own new species, just as a Victorian primary school girl and two university students have already done!

Figure 9.15 Sometimes fossils and equipment have been flown out of Dinosaur Cove aboard helicopters provided by the Surf Lifesaving Association of Victoria and the Victorian Police. More often they are carried up the 80 to 100 metre cliff on someone's back or are sent up on the flying fox. (F. Stewart)

DRIFTING CONTINENTS AND POLAR DINOSAURS

Australia not only has its own unique species of dinosaurs but it also has *polar* dinosaurs. There is only one other place in the world where many polar dinosaur bones are known and this is on the North Slope of Alaska, also just recently discovered. So, there is a fair bit of interest in working out just what kind of place Australia, and even more specificially, southeastern Australia was like 105 to 120 million years ago.

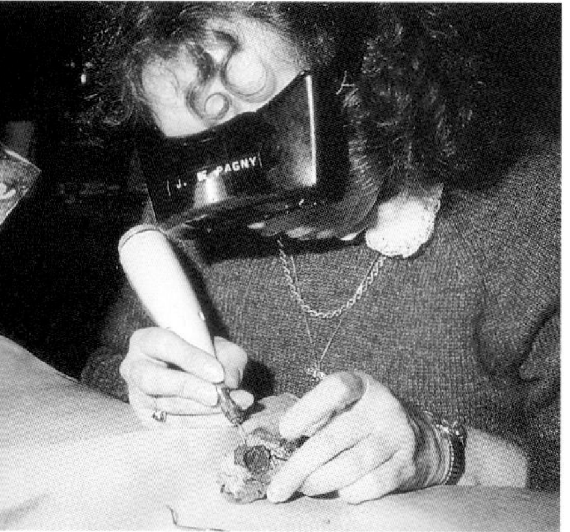

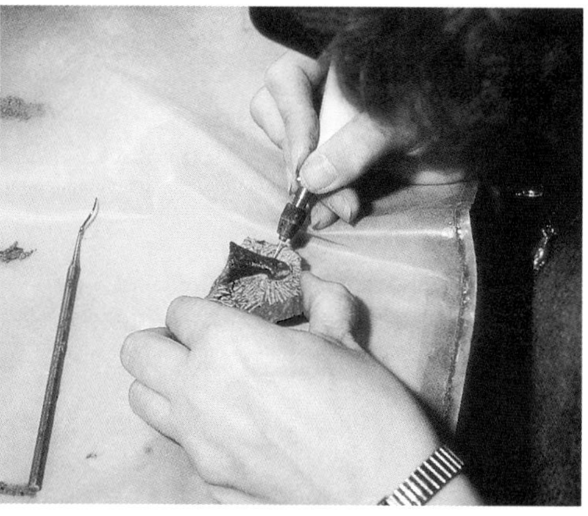

Figure 9.16. The preparation of fossils from Dinosaur Cove requires much patience, small needles, electrically driven micro-jackhammers - *and* a long time. Some fossils have taken weeks, even months, to prepare. (F. Coffa)

Many people want to know just how dinosaurs managed to cope with the cold polar conditions - in which a dark winter with no sun at all lasted for 3 solid months, and where there was also a 3 month long summer, with sunlight 24 hours a day. Did the dinosaurs simply migrate out of the region to find the light during the winter, or did they stay and tough it out, either by hibernating or continuing to operate even during the coldest times? Certain animals such as musk ox and caribou can survive such winter conditions nowadays - but they are warm-blooded mammals and have a fur coat to insulate their bodies. So, were the dinosaurs that lived in Victoria, the chicken-sized to human-sized hypsilophodonts and the tiny 3 metre high carnivorous *Allosaurus*, warm blooded too?

Figure 9.17. Small dinosaur jaw (about 5 cm long) in the rock, Dinosaur Cove. (F. Coffa)

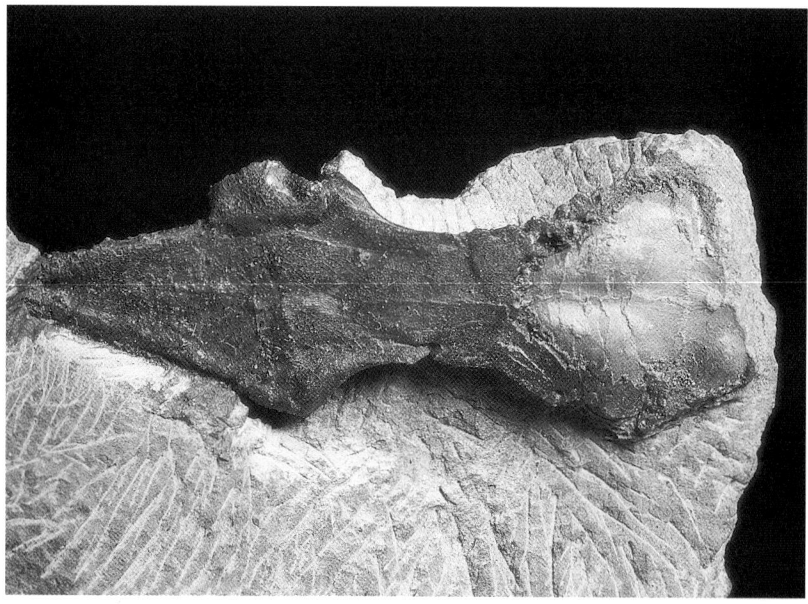

Figure 9.18. Results of preparation can be spectacular. This specimen of a tiny hypsilophodontid dinosaur no larger than a bantam hen, shows the impression of its brain on the right side of the picture. (S. Morton)

Figure 9.19. The astragalus, anklebone of an *Allosaurus*, a 4-metre-high flesh-eating dinosaur. This bone was found along the coast near Inverloch, Victoria in Cretaceous rocks slightly older than those at Dinosaur Cove. (F. Coffa)

The hypsilophodonts that lived here certainly had large brains for the average dinosaur, and also very large eyes. Were these adaptations for surviving under very low light conditions? We do know that certain types of birds that feed mainly at dusk and dawn have much larger eyes than their day feeding 'cousins.' Our polar dinosaurs are fascinating, but they still hold many secrets that we have yet to uncover as we persist in excavating their 110 million year old tombs

There are many easier places to collect dinosaurs in Australia, but easier though they may be, they do not produce the bones with such regularity as the ancient river and lake rocks that are exposed along the south coast of Victoria.

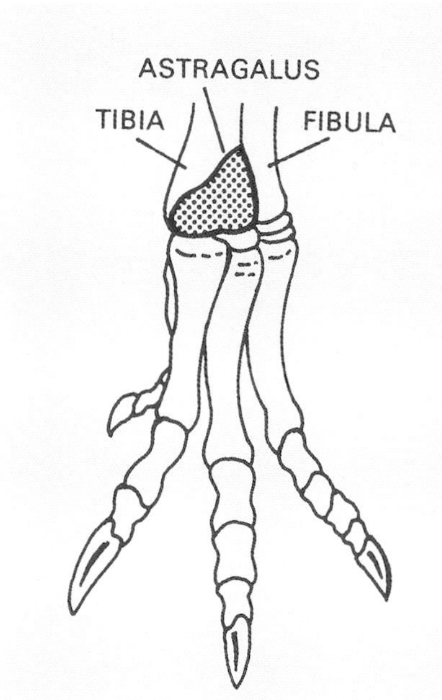

Figure 9.20. Drawing of an anklebone (astragalus) of an *Allosaurus*.

	W OTWAY BASIN										GIPPSLAND BASIN					E
LOCALITY / TAXA	Casterton	Dinosaur Cove West	Dinosaur Cove East	Slippery Rock	Knowledge Creek	Eric the Red	Point Franklin	Point Lewis	Marengo	Cumberland River	Punch Bowl	Kilcunda	Eagles Nest	Cape Paterson	Koonwarra	Wonthaggi
Dinosaur indet.		O	O	O		O	O	O	O		O		O			
Fulgurotherium australe			O									O	O			
Victorian hypsilophodont A		O	O	O												
Victorian hypsilophodont B		O	O	O				O								
Victorian hypsilophodont C			O	O												
Victorian hypsilophodont D														O		
Theropoda								?	O							
Allosaurus sp.													O			
Dinosaur footprint					O											
Testudines		O	O	O									O			O
Chelycarapookus arcuatus	O															
Plesiosaur		O		O						O			?			
?Lepidosaur													O			
Pterosaur			O	?												
Aves															O	
Labyrinthodont											O					
Pisces				O												
Ceratodontidae				O												
Ceratodus avus													O			O
Ceratodus nargun								O								
Ceratodus sp.															O	
Coccolepis woodwardi															O	
Wadeicthys oxyops															O	
Koonwarria manifrons															O	
Leptolepis koonwarri															O	

Figure 9.21. Fossil vertebrates found at Dinosaur Cove, Victoria, from the early Cretaceous.

QUEENSLAND'S DINOSAURS

DINOSAURS NEARER THE EQUATOR

The dry and barren wastes of western Queensland have produced the largest bones of Australian dinosaurs and even a few partial or nearly complete skeletons. *Rhoetosaurus*, a large sauropod with four feet firmly planted on the ground, is Australia's biggest dinosaur. Some of its bones were first dug up in 1924 on Durham Downs Station, near the town of Roma in Queensland. Eventually some of the bones were sent to the Queensland Museum in Brisbane - and there they were correctly recognised as parts of a gigantic dinosaur skeleton. More bits were collected over the next few months, and finally the director of the museum wrote a paper on them - announcing them to the scientific world. He gave this new dinosaur the name *Rhoetosaurus* - after the giant of Greek myths *Rhoetos* combined with the standard ending used on dinosaur names, *saurus*, meaning lizard in Latin - 'giant lizard'. Bones of this 12-metre-or-so long beast were found over the next 50 years, but nothing very complete came to light. In 1975, however, a most determined curator at the Queensland Museum, Mary Wade, thought that, based on the bones in the museum, there was a good chance a whole skeleton lay buried near Roma. The chase was on - Mary organised an expedition to the area, searched out the sites that had already yielded bones, and started to dig at the most likely spot - the result was several more parts of the skeleton that had provided the bones lying in the Brisbane Museum, including an almost complete hind foot.

All these years of chance finds and Mary's expedition gave Australia its only glimpse of a complete home-grown sauropod - part of the neck, back, pelvis, ribs and hind legs as well as a fair bit of its long tail. Although the actual collection of *Rhoetosaurus* may not have been as difficult as the operations at Dinosaur Cove, discovery was not easy. Bones were locked in a tough sandstone, just as at Dinosaur Cove. So, even though the wind and weather had exposed more of the bones, once back in Brisbane, the people (preparators) chipping away the rock with miniature jackhammers, called vibratools, or sandblasters, had a hard job freeing such fossils from their rocky tombs.

The Queensland Museum is the home of the bones of *Rhoetosaurus*, and it is also the place in Australia where you can see the most complete skeleton of an Australian dinosaur - *Muttaburrasaurus*. You don't even have to try to imagine what it looked like; it's just about all there - just inside the front doors of the museum. Its skeleton, standing on its two hind legs, makes you look very small and gives you an immediate impression of the size of this plant-eating dinosaur. The first of its bones came to light in 1963 during a cattle muster on the banks of the Thomson River near Muttaburra, to the northeast of Longreach in western Queensland. Many people in the area had collected parts of the skeleton and taken them home, but it wasn't until a bit later that the Queensland Museum found out about the dinosaur. When the palaeontologists from the museum visited the area and put out a public appeal to return bones, the response was excellent - bones were taken off the

Figure 9.22. Some of the dinosaurs of Queensland were generally much larger than those of Victoria. Included here are the big sauropod *Rhoetosaurus*, the iguanodont *Muttaburrasaurus*, the armoured dinosaur *Minmi*, small- insect or small-reptile-eating coelurosaurs, the medium-sized hypsilophodonts and the carnivorous theropods. (Painting by M. Hallett, courtesy of the Queensland Museum).

kitchen window sills and mantlepieces and given to the museum. And, as a result, it was possible to reconstruct most of the skeleton. Each of those Queenslanders who donated their prehistoric relics can pat themselves on the back every time they visit the Queensland Museum, because they played a very important part in helping piece together Australia's most complete and unique dinosaur.

Muttaburrasaurus was a terrestrial animal, that is, it lived on land. But, quite interestingly, its bones were found in rocks deposited by the sea, a shallow ocean that had flooded onto the Australian continent, dividing it into many small islands. Evidently this *Muttaburrasaurus* had died and its rotting carcass had floated down a stream, (probably not too far, because the skeleton was still nearly complete), and had been dumped where the river entered the sea. We know that the rocks containing the dinosaur skeleton are from a marine environment because the shells of marine animals are preserved in them - the fossils of ammonites (relatives of the squid and octopus), clams, and snails, all of which lived only in the sea.

Figure 9.23. *Rhoetosaurus*, a sauropod dinosaur from the Jurassic of Queensland. (Drawing by F. Knight, courtesy of the Museum of Victoria)

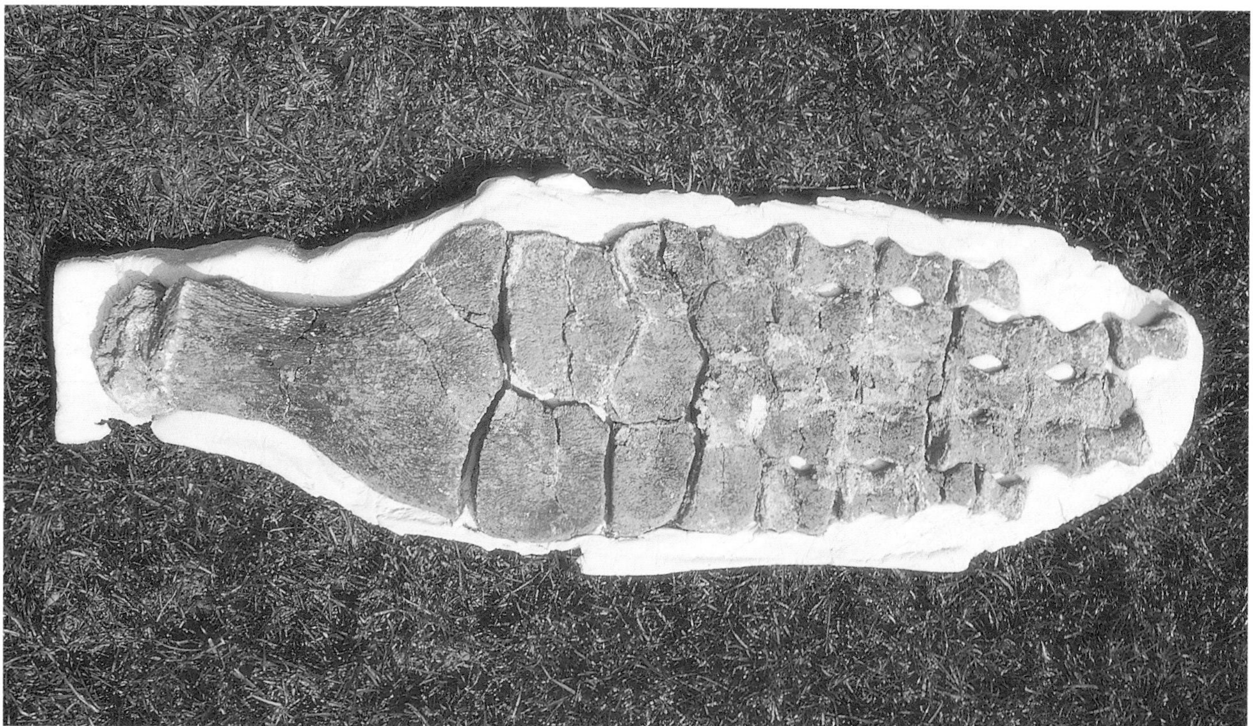

Figure 9.24. Paddle of a plesiosaur, a Loch-Ness-monster-like marine reptile that swam in the Cretaceous seas of Australia.

In that same sea, some 100 million years ago lived some of the most amazing sea monsters, the plesiosaurs, which were Australia's answer to the Loch Ness Monster, and the ichthyosaurs, dolphin-like reptiles that probably gave birth to their young alive rather than laying eggs. These reptiles were streamlined like submarines, fast moving predators of the depths and shallows, as dangerous as sharks to all that lived in their seas.

The sea that received the *Muttaburrasaurus* skeleton and that also buried the bones of the big reptile carnivores was not a warm tropical one that you might expect. Professor Larry Frakes and those working with him at the University of Adelaide think that they have fairly good evidence that the sea had icebergs in it at times - at least in the area around northern South Australia. They have found big rocks, which they call dropstones, in a very fine sediment called the Bulldog Shale (actually petrified mud). "How did those rocks get in that mud?" they asked themselves. One sensible explanation that they they came up with was that icebergs had floated out into the sea carrying rocks they had picked up as they moved along the ground when the ice was part of a glacier. When that glacier met the sea and icebergs were born (or calved), they carried their rocky baggage with them. Then they floated out into the sea, where the plesiosaurs and ichthyosaurs swam. The icebergs melted and dropped the rocks, which fell into the muds on the sea bottom....an environment far from a tropical paradise! Now, we are beginning to wonder if the plesiosaurs and ichthyosaurs that lived in the cool to cold oceans were warm-blooded too. So far we don't know the answer, but maybe someday we will. It is these interesting, but not yet answered questions that make science worth pursuing!

Figure 9.25. *Platypterygius*, a dolphin-like Cretaceous reptile, an ichthyosaur that swam in the seas flooding onto the Australian continent. (Drawing by F. Knight, courtesy of the Museum of Victoria)

Figure 9.26. A Cretaceous plesiosaur. Mainly marine reptiles, some freshwater-dwelling are known from Dinosaur Cove. (Drawing by F. Knight, courtesy of the Museum of Victoria, from Rich, van Tets and Knight, 1985)

Figure 9.27. Above and in the shallow cool seas that flooded western Queensland in the Cretaceous lived *Ornithocheirus*, a flying reptile, and *Notochelone*, a marine turtle swam below. (Drawing by F. Knight, courtesy of the Museum of Victoria)

EXTINCTION OF THE DINOSAURS

EXTINCTION, THE FINAL DAYS, WHY ?

If you had been on a holiday in Australia 70 million years ago, you most surely would have crossed paths with a dinosaur somewhere. But if you had decided to return about 60 million years ago, you would have been, for the most part, out of luck. By this time, in the early part of the Cainozoic Era, in the Palaeocene Period to be more exact, dinosaurs had disappeared. Why? There are many theories, but lots of people are still arguing about which one is right. There are two basic ideas about why a group that lived so long, for about 140 million years, finally died out. One theory says that the earth was hit by a big comet or meteor, or several. This did a few strange things to the environments at the time. Over a short period of time it probably led to the drastic cooling of the climate as a lot of dust was thrown up into the atmosphere, and this cut down the incoming energy from the sun and thus lowered temperatures on a world-wide scale. It got cold, and animals and plants that couldn't cope with that died out — dinosaurs (all except birds, which are living dinosaurs!), ammonites, and sea monsters! Before the cooling of the atmosphere, which probably took a while, there may have been a period of intense heat that occurred just as the earth and this unwelcome extra-terrestrial visitor collided. When the comet or meteor disintegrated, it might well have produced an acid rain on a massive scale as well as possibly introducing a number of unusual poisons into the earth's atmosphere. All these factors could have helped knock out a number of different animal groups. Perhaps even the prolonged darkness that might have come about as a result of all the dust in the air could have made life very difficult for animals that were not warm-blooded and had no ability to hibernate. These are all possibilities of the theory and there are many more.

The second extinction theory says that climatic change was *the* important cause of the extinction of most of the dinosaurs and lots of other Mesozoic life. The climatic change was one of cooling, the cause of which may have been related to the beginning of the breakup of the continents. When the big supercontinents separated, the arrangement of the oceans changed. With that change there came a change in ocean circulation; that is, the location of its major currents changed. Thus, the distribution of energy from equator to pole and pole to equator changed. The climate may have become cooler. Ice caps, which seem to have been non-existent in the Cretaceous, began to build up, and the world began to take on its present-day climatic form. Maybe it was this cooling that led to the death of the dinosaurs and many of their "friends".

So, which theory is best? As we mentioned above, that is another unresolved question. Recently a group of geologists have suggested than an increase in volcanic activity - lots of volcanoes erupting at about the same time, may be just as good or better an explanation for the extinction of the dinosaurs as the crash of a meteor. Lots of volcanic activity would be expected during times of increased rate of drift of the

Figure 9.28. Was the extinction of the dinosaurs caused by the collision with the earth of a meterorite like this one? (F. Coffa)

Figure 9.29. A crater in Arizona USA that resulted from the collision of a meteor with the earth (W. Birch)

continents, which we know was happening at the end of the Mesozoic. The two phenomena are related because as a new ocean basin forms, lots of volcanic lava flows into the ocean basins and essentially pushes the continents apart. As the big slab of continental crust plunges down into oceanic trenches, like those under Japan today, this causes volcanic eruptions. If the continental movement accelerates, the trench related volcanic activity will also increase. Volcanoes throw out a lot of ash and rock particles into the air - and these, then, form dust clouds that can essentially cut down on the amount of light that reaches the earth's surface from the sun. We know about this by looking at the effects that major volcanic eruptions have had on climate over the last century. Perhaps volcanoes could have had just as much an effect as a big meteor - and maybe it was the volcanoes that blasted the dinosaurs out of existence, so to speak. Some geologists have recently suggested that meteor impacts could have set off massive volcanic activity and together caused the mass extinction 65 million years ago. We'll have to do a great deal more scientific sleuthing before we can work that one out, if, in fact, we ever do!

There is no doubt at all, however, that once the dinosaurs were out of the picture, the stage was set for the remarkable success of the mammals, and eventually, us humans.

Figure 9.30. Scientists digging up dinosaurs are not the only people fascinated with dinosaur bones. The press and many people who read the papers and watch the news are also very interested. From studying what happened to the dinosaurs 65 million years we may learn something about our own future. But unlike dinosaurs, we have the ability to change things. Whether we do that sensibly and with foresight or stupidly and without thought of consequence, will determine our eventual destiny. (F. Coffa)

The Jurassic World
208 to 144 million years ago

Australia really has very few dinosaurs of Jurassic age. Part of the continent lay above sea level at this time. Climates were warmer and wetter than in the past. The vegetation was rich with conifer forests and cycads, seed ferns, and a variety of ferns. Horsetails lived in the coal forming swamps. Freshwater fish swam in the broad rivers, which flowed over much of eastern Australia, and labyrinthodont amphibians and dinosaurs walked the dry land and its swamps.

Figure 9.31. A family tree of Mesozoic reptiles - how dinosaurs are related to other living things.

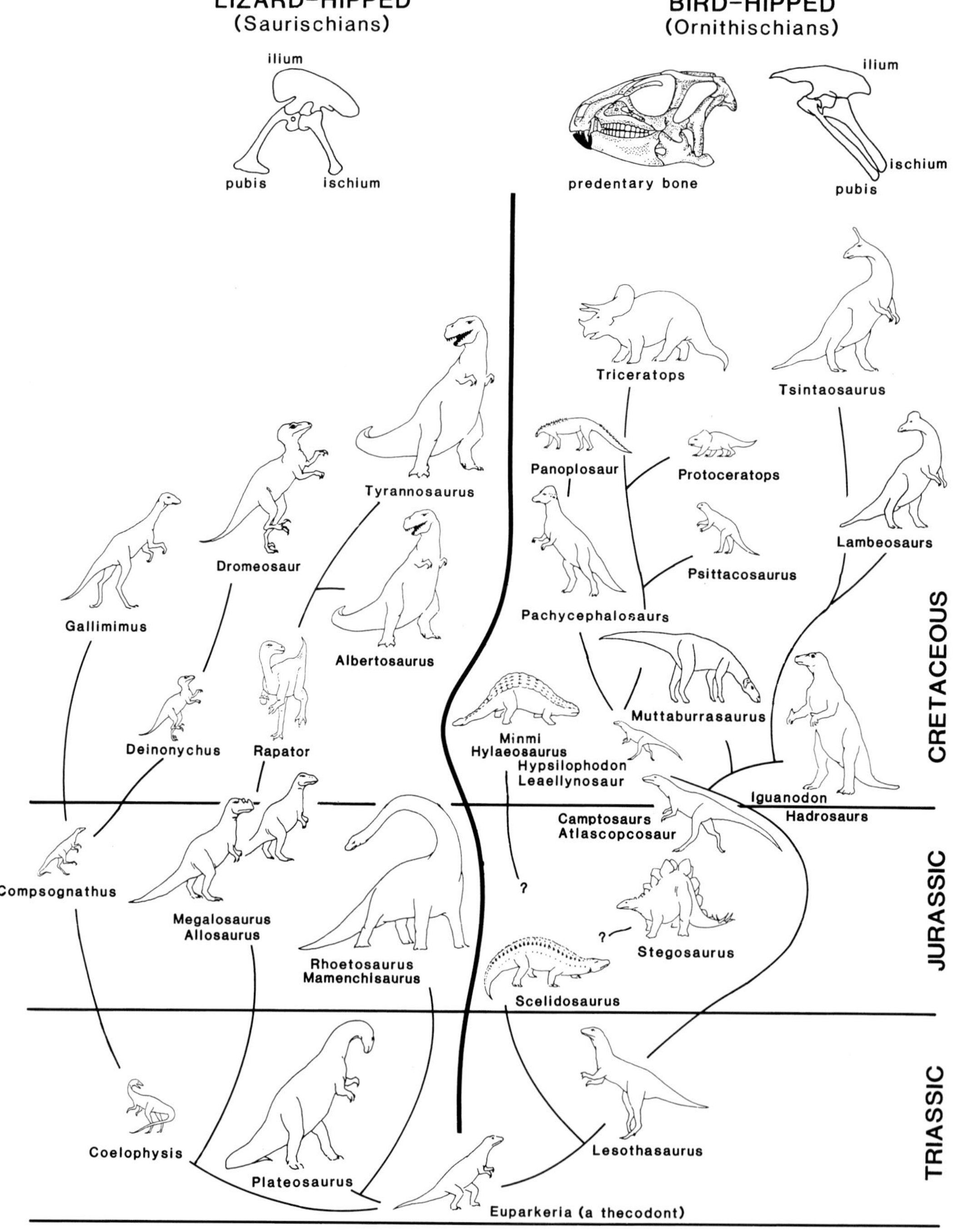

A DINOSAUR'S ENVIRONMENT

Figure 9.32. The skull of *Tyrannosaurus*, a big carnivorous dinosaur related to *Allosaurus*. The dagger-like teeth were serrated like a steak-knife and quite able to slice flesh into small pieces, ready for further processing by the gut. (F. Coffa)

The Environment

For much of the Jurassic, Australia was above sea level. Early in the Jurassic, earth movements produced a deep, narrow trough in the west that filled first with freshwater sediments. Later this area was invaded by the sea. In the east, a large river system developed, which gradually changed to coal swamps and freshwater lagoons. Volcanoes erupted along the south coast, and rivers and lakes developed around these volcanic mountains.

The Plants

The plants of Jurassic Australia were like those present in most of the world at this time, a mixture of ancient and modern conifers, cycads, seed-ferns, ginkgos, ferns, and low growing club mosses and horsetails. The plants lived on into the early Cretaceous, and many then became extinct, the last time a really ancient plant community lived in Australia.

Coal was deposited, which means there were large areas that were quite wet, but there were arid areas, too, in Australia's southwest.

Plants from the middle Jurassic Talbragar area of central New South Wales give a good idea of the plant communities of these times. A Kauri Pine (*Agathis*) forest grew around a large lake there. Sharing the forest with *Agathis* were podocarpaceous conifers and the cycad *Pentoxylon* which formed the lower growing plants. Heathy areas around lake shores were covered with seed ferns. Horsetails lived in the swampy areas.

Backboned Animals

Jurassic vertebrates lived both on land and in the water. A variety of dinosaurs such as the large theropod, *Changpeipus*, left only their footprints in the Walloon Coal Measures of

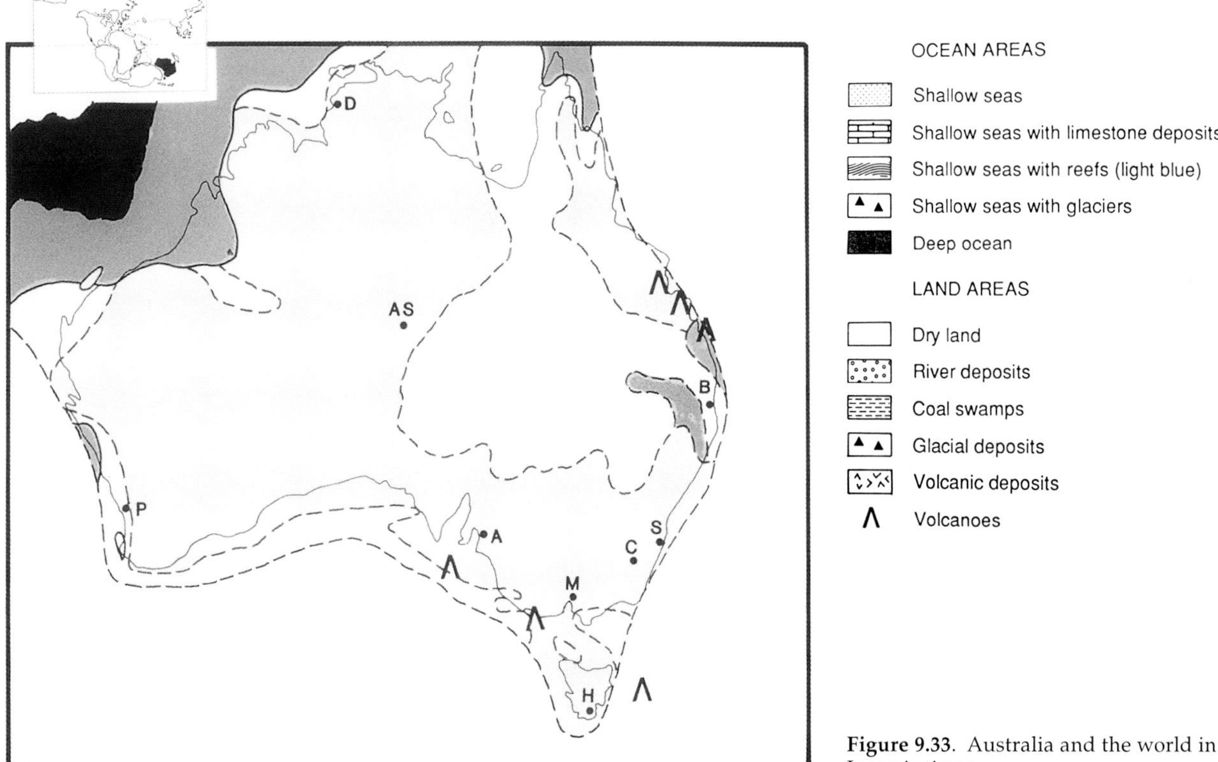

Figure 9.33. Australia and the world in Jurassic times.

Queensland. Both large and small carnivorous theropods were present, as was a larger sauropod, *Rhoetosaurus*, which left a partial skeleton on a Queensland floodplain. Labyrinthodont amphibians (*Siderops*) lived into the early Jurassic, one of the youngest occurrences of this group in the world. Marine waters were ruled by the reptilian "sea monsters" plesiosaurs.

Fish were abundant, as the mid-Jurassic Talbragar fossils illustrate. The fish faunas were dominated by ray-finned forms both primitive (*Coccolepis*) and advanced (*Leptolepis*). Teleosts, advanced ray-finned fish, are the dominant fish of today, our trout and barramundi. So, even in the Jurassic, a glimmer of the modern world could be seen.

THE CRETACEOUS WORLD
144 to 65 million years ago

Australia may have moved a little closer to the South Pole in the Cretaceous. Climates were seasonal and probably cool, at least along the southern parts of the continent where there was a three-month-long polar night each year. Early in the Cretaceous, much of the continent was invaded by a wide, shallow sea, which drained off at the end of the period. Vegetation was dominated by ferns, cycads, ginkgos, and conifers, but the first flowering plants also appeared. Plesiosaurs and ichthyosaurs, together with many different kinds of bony fish, lived in the shallow seas. On land, dinosaurs ruled, but the last surviving labyrinthodont amphibians anywhere on earth finally died out in the early Cretaceous, and the first Australian mammals and birds made their appearance.

Figure 9.34. (*Right above*) The auracarian conifer was one of the main trees in Cretaceous forests. Today these plants are not common, but the Norfolk Island Pine is one that has survived. You can buy such trees now in most nurseries and so own your own dinosaur tree!

Figure 9.35. (*Right below*) The ginkgo or Maiden Hair Tree was thought to be extinct until it was found growing in a monastery in China not many decades ago. It was widespread and quite abundant in the Cretaceous of Victoria. Female trees produce a terrible smelling fruit. Mammals, like us, hate the smell, but dinosaurs may have loved it. (F. Coffa)

Figure 9.36. These leaves belong to an auracarian conifer, rare now, but abundant during the reign of dinosaurs in Victoria. (F. Coffa)

Figure 9.37. Food fit for dinosaurs. Fragments of leaves from ferns and conifers from the Boola Boola area of Victoria, early Cretaceous in age. (S. Morton)

Figure 9.38. (*Lower left*) The spore of a lycopod, a club moss from the early Cretaceous rocks of southern Victoria. (B. Wagstaff, J. McEwen-Mason)

Figure 9.39. (*Lower centre*) A fern spore from the early Cretaceous rocks of southern Victoria. To be studied, these tiny fossils must be magnified about 500 to 1000 times their natural size. Ferns were very abundant in Victoria 106 million years ago. (B. Wagstaff)

Figure 9.39a. (*Lower right*) A living fern in Australia survives quite well even in the snow – perhaps like those when dinosaurs lived in Southern Victoria.

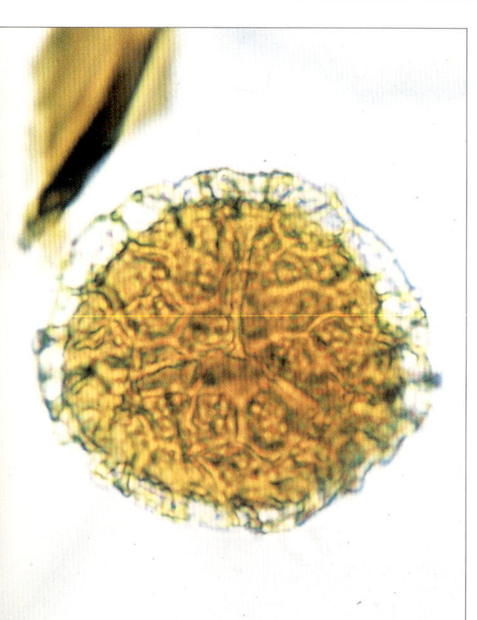

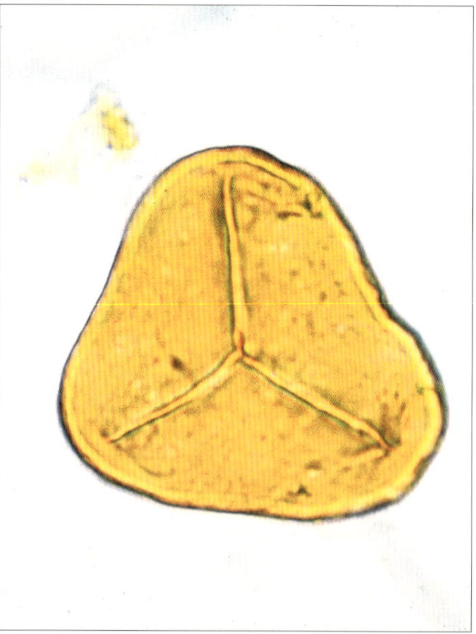

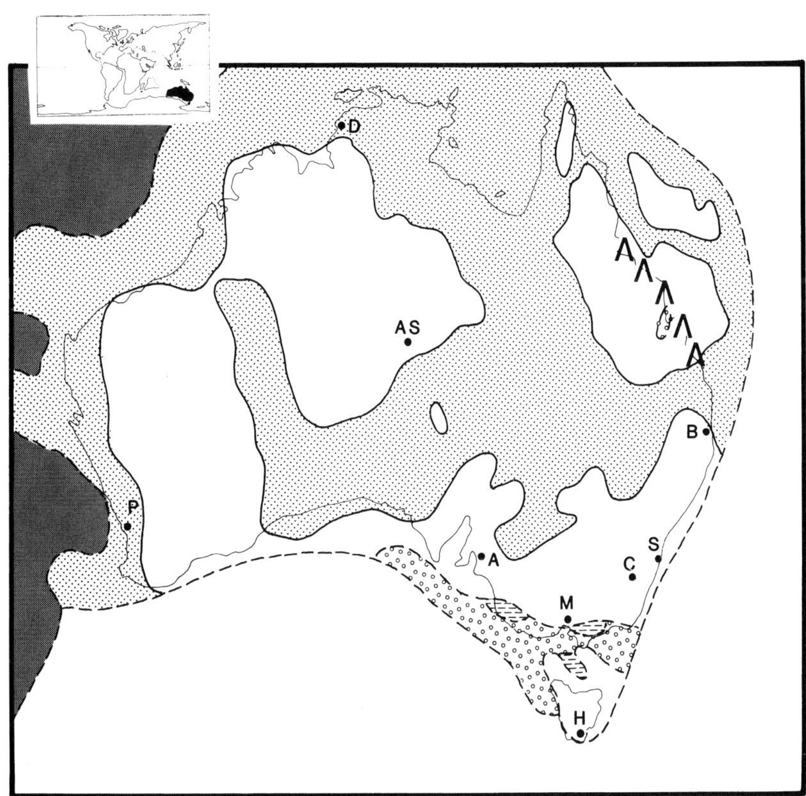

Figure 9.40. Australia and the world in Jurassic and Cretaceous times.

The Environment

Gondwana continued to break apart in the Cretaceous, with the south coast of Australia facing a gradually widening rift valley. Volcanoes to the east showered ash and debris in that valley dotted with lakes and large, wandering rivers. Streams flowing off the cold mountains brought icy waters into the valleys below. Australia still was near the South Pole through all of this period.

In the early Cretaceous, as global temperatures rose, sea floor spreading increased, sea levels rose, and oceans flooded onto much of Australia, dividing it into several big islands. The western Pacific Ocean was wide and warm, and ocean currents flowed towards the poles, bringing warm and wet conditions to high latitudes, certainly not what happens today in this region.

In the late Cretaceous, the whole world began to cool but because Australia began to move north, away from Antarctica, its temperatures actually rose. Shallow seas drained off Australia, as more and more water was frozen into ice. Broad rivers flowed across Queensland, depositing the Winton Formation, which preserved a variety of vertebrates.

The Plants

The early Cretaceous plants of Australia were similar to those of the Jurassic, with forests of conifers such as podocarps and araucarians (Norfolk Island Pines, for example) along with ginkgoes forming the forest trees. Seed ferns, ferns, and cycads lived on the forest floors. Then changes began, with startling suddeness. Flowering plants, the angiosperms, appeared at the end of the early Cretaceous.

Pollen and leaves of these flowering plants are known in Australian rocks. Angiosperms may have first lived in the disturbed coastal valleys that developed as Australia broke away from the rest of Gondwana and elsewhere in the big rifts that were splitting continents apart at this time too. Initially small, these "weeds" developed into groups of new species, and by the end of the Cretaceous had crowded out many of the conifers. Modern Australian flowering plants that also lived in late Cretaceous times include the Proteaceae (waratahs and their relatives), and *Nothofagus*, the Southern Beeches. Plants began to look very modern.

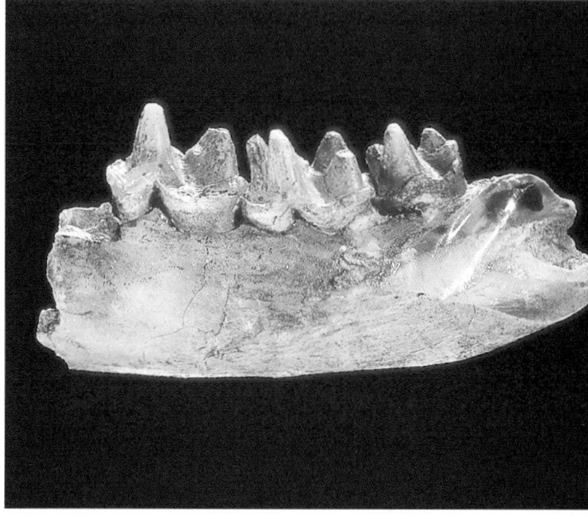

Figure 9.41. The jaw of one of Australia's rare Cretaceous mammals, *Steropodon*, a toothed platypus. This beautiful opalised specimen was found by an opal miner in 115 million year old rocks at Lightning Ridge, New South Wales, and was first recognised by Alex Ritchie, Curator of Fossils at the Australian Museum. Alex recognised the jaw when it was poured out on a hotel bed with many other fossil bones by a far-sighted opal-dealer, who for years had been buying fossils from the opal-miners and then bringing them to the museum curator for examination. (Courtesy of the Australian Museum, Sydney, J. Fields and A. Ritchie)

The Backboned Animals

Land-dwelling vertebrates were quite varied in Cretaceous times, but fossils of them are not preserved in great quantities. The first Australian mammals (platypuses with teeth) appeared, as did birds, the only traces of which are their fossilized feathers at one place along the south coast of Victoria, as well as three tiny bones in Queensland.

Several different types of dinosaurs lived in the river valleys of Victoria and Queensland. The southern Australian dinosaurs were small, mainly hypsilophodontids, dinosaurs which are generally rare at more tropical latitudes. Even the carnivorous theropods of these polar areas were small. In Queensland, larger forms, such as the big ornithopod *Muttaburrasaurus* and the armoured *Minmi*, as well as sauropods, were alive and well. Flying reptiles, such as *Ornithocheirus*, shared the skies with birds, and the freshwaters had several kinds of turtles and fish, such as *Ceratodus* and the present-day *Neoceratodus*, both lungfish, and many ray-finned forms including both primitive and advanced types. The last Australian palaeoniscid, *Coccolepis*, a very primitive ray-finned fish, lived his or her final days in Cretaceous rivers. Freshwater insects, spiders and many other invertebrates inhabited one early Cretaceous lake at Koonwarra, Victoria whose cool temperate conditions were preferred by the animals.

Figure 9.42. Several femora (thigh bones) of three to five different kinds of small hypsilophodont dinosaurs. Babies and adults are represented, and one was larger than a medium-sized human. These come from the early Cretaceous rocks of southern Victoria, 106-115 million years old. (F. Coffa)

The oceans were full of several kinds of marine reptiles - the dolphin-like ichthyosaurs (*Platypterygius*), Loch Ness Monster-like plesiosaurs (*Kronosaurus, Woolungasaurus*), and the fish-lizards, the mosasaurs. Fish, such as sharks (*Carcharias*), ray-finned forms (*Belonostomus*) and the advanced ray-finned fish, the teleosts (*Xiphactinus* and *Pachyrhizodus*), probably were food to the big reptiles. The shallow seas in which these animals lived were probably quite cool.

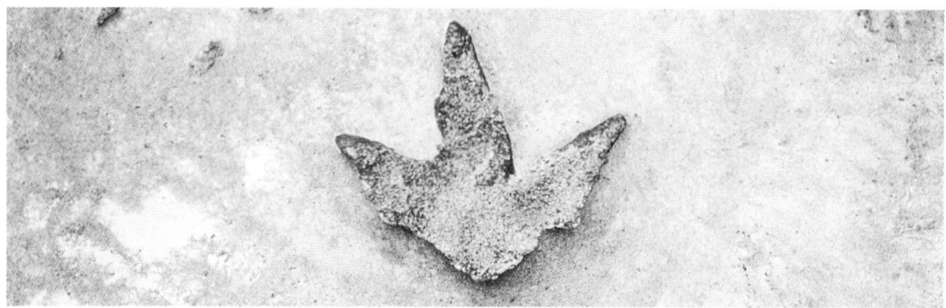

Figure 9.43. The footprint of a small carnivorous dinosaur that ran on its hind legs. This is one of three footprints of dinosaurs known in Victoria, and proves that dinosaurs walked or ran across the river plains of this area about 106 million years ago. (F. Coffa)

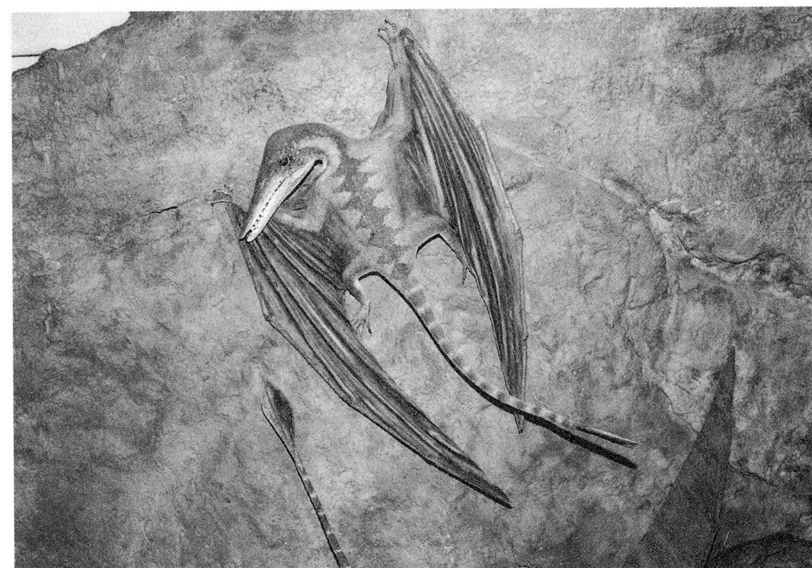

Figure 9.44. A flying reptile, a pterosaur. Two bones of this reptile group have been found in the early Cretaceous rocks of Victoria. These were the bats of the time. Real mammalian bats didn't appear for nearly another 50 million years. (F. Coffa)

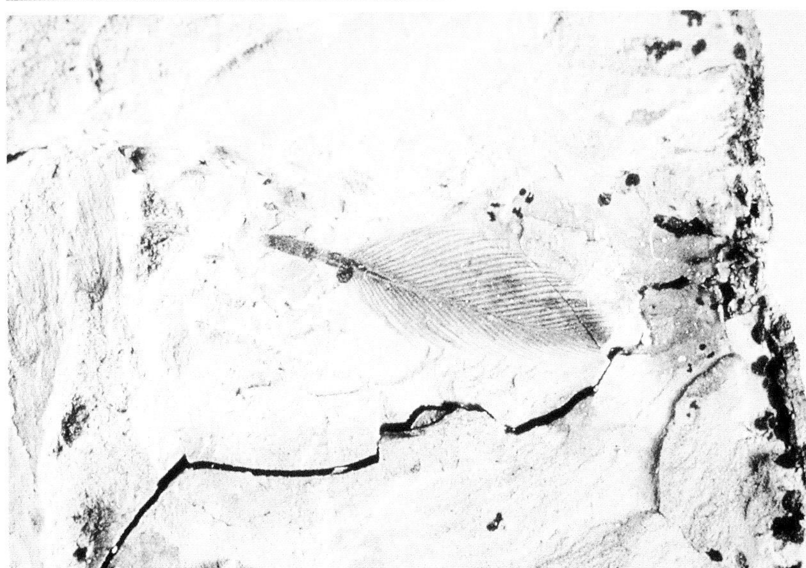

Figure 9.45. The feather of a bird, from Cretaceous rocks of Koonwarra, Victoria. (F. Coffa)

Figure 9.46. A fossil bony fish from Koonwarra, an early Cretaceous lake deposit in South Gippsland, Victoria. (F. Coffa)

Figure 9.47. *Minmi*, an armoured dinosaur from the Cretaceous of Queensland. (Drawing by F. Knight, courtesy of the Museum of Victoria)

CHAPTER 10:
RAINFORESTS TURN TO DUST

Australia Parts Company with Antarctica: Cool, Wet Forests, and Flamingo Covered Lakes of the Centre. The Great Drying Out, Our Deserts Begin

The Cainozoic
65 Million Years Ago to Now

We had been churning slowly along the Birdsville Track north of Maree, in and out of big mud puddles. It had rained heavily the night before as we lay in our beds in the Great Northern Hotel, our last indoor bed for two months - and our last reliable, long hot shower. We wondered just how long the rain would go on and what the track would look like in the morning. The next mudhole brought orangy-brown water across the floorboards in our little Land Rover, but had no effect on the motor, which just kept going - sounded under little strain, so the bottoms of these deep puddles were still solid - it was just that the water was deep. At about three o'clock we reached the artesian bore that marked the spot where we turned off to head several kilometres cross-country to one of the usually dry lakes where fossil vertebrates had been found. We stopped and looked out across the wet gibber plain - it looked awfully muddy. The main problem was the bore drain. Once we got across that, the next few kilometres, although slippery, would probably be OK! There were three vehicles, a Land Rover, a Nissan, and a Toyota. The 'Landy' went first, in four-wheel drive, low range - in 2nd, then 3rd gear. We slipped and slid about, and the motor groaned - but in three to four minutes we had crossed the few hundred metres of difficult country and gained the high ground out towards Lake Palankarinna.

We made it without too much trouble but had left behind some rather deep ruts - not too helpful for the other two vehicles that were to follow. Neville got through in his Nissan with a bit more strife, at one time nearly clearing the ground with all four wheels, but the Toyota became hopelessly bogged halfway across the rutted track we had left behind. Mike looked disgusted. It was our fault, not his - our ruts had caused him to get stuck, so

Figure 10.1. Most dinosaurs were extinct by the end of the Cretaceous (except for birds), along with a great variety of other life forms, such as marine plankton. Mammals and birds began to multiply rapidly in the early part of the Cainozoic. (From *The Tower of Time*, a mural by John Gurche, courtesy of the Smithsonian Institution, Washington, D.C.)

Figure 10.2. Bogged!

Figure 10.3. The Great Northern Hotel in Marree— last hot shower before weeks in the field.

he quite rightfully waited for Neville to slosh back through the sticky puddled mud with a winch cable that quickly pulled the Toyota free. By sunset we had all reached our campsite near a big sand dune. We set about putting our tents up and the billy on for a welcome cup of tea, with a dash of rum to warm the bones. The bogging was quickly forgotten.

FROM THE WET TO THE DRY

PLANTS AND ENVIRONMENTS

Although rain in central Australia can quickly turn parts of it into a mire or temporarily fill the thousands of dry salt pans, for the most part, for most years, it is truly a desert, a very dry place. It has not been that way forever. In fact, it hasn't been that way for very long - less than a million years.

A million years seems a long time. In terms of a human lifetime, it is a long time, but as we saw in Chapter 2, when compared to the 4600 million year history of our planet and the more than 3500 million year history of life on the Australian continent, a million years is hardly anything at all! Today's Australia, with its deserts and only a few spots of rainforest, with its tiny pockets of cycads such as those of Palm Valley in the Northern Territory, is a very recent development.

The last 65 million years in Australia have been times of rather drastic changes in geography and climate, changes that are quite readily shown by the fossil plants and animals that we know from those times.

Figure 10.4. A 15-million-year-old, possibly older, tree trunk weathering out on the mud cracked surface of an unnamed lake near Lake Frome, South Australia. Where desert rules today, great forests once grew along lazy streams, and lakes were filled with flamingoes and dolphins – even a stray platypus, with teeth! Today that is all changed.

Figure 10.5. Central Australia today – far from a rainforest.

Figure 10.6. The Sturt Desert Pea, a native Australian angiosperm that thrives in the central Australian deserts.

RAINFORESTS IN CENTRAL AUSTRALIA

At the beginning of Cainozoic times, Australia was still quite firmly attached to Antarctica, and lay far south. Parts of southern Australia still had a winter with at least three months of continuous darkness, lit only by the moon, the stars, and the Aurora Australis or Southern Lights.

Fossilised pollen grains trapped in the sediments as silts and clays that were laid down in Cainozoic lakes and swamps give a clue to what the plants were like and so what the climate was like. Some places have also preserved leaves, twigs, and even fruits from this time. Especially interesting are localities in the now dry centre of Australia. Understanding what has happened there over the last 50 to 60 million years will give clues as to how and when the deserts came into existence.

One area to the southwest of Lake Eyre (not often a lake with water but generally a dry saltpan) has produced fossils of fruits and leaves - of primitive conifers (araucarians and podocarps) as well as of gums, paperbarks (*Melaleuca*) and bottlebrushes (*Callistemon*). These are thought to be of early Tertiary age, perhaps around 40 to 50 million years old. Fossils of pollen and spores from the same areas, as well as from other places in central Australia, have been recovered from the drilling of wells. These tell much the same story. During the early part of the Cainozoic, up until the Oligocene, a large part of what is now arid country in Australia was much wetter. Pollen of *Dacrydium*, primitive conifers, ferns, and of flowering plants whose relatives now live in areas of high rainfall, such as *Nothofagus*, the southern beeches, are the ones that are most abundant fossils. The plant fossils show that cool, wet rainforests existed in many parts of central Australia where only sand dunes and spinifex are now to be found.

Figure 10.7. Fossilised leaves of angiosperms from Cainozoic rocks in Australia show that the flowering plants came to rule this time, pushing out more ancient plants and even outdoing the conifers. (S. Morton)

Figure 10.8. Wet rainforests ruled much of Australia in the first half of the Cainozoic.

We also know from looking at the sediments that were laid down at this time that rivers and lakes were scattered all over South Australia, Western Australia and the Northern Territory - in fact, over most of central Australia. These were permanent and their existence indicates that lots of water was flowing or collecting in them. They weren't like the rivers there today, like Cooper Creek, that flow only once every 10 or 20 years!

Rainforests didn't cover all of central Australia, however. The first grasses appeared in the Eocene, as did pollen of plants related to the gums, so there must have been open, drier country as well. Still, rainforests ruled, certainly until 20 million or so years ago. Temperatures fluctuated but may at times have been as high as 5 degrees Celsius warmer than now. There were probably also times, however, such as about 30 to 35 million years ago, when temperatures were just about like they are now. Temperatures have been figured out by geochemists working on the oxygen isotope content in fossil shells in marine rocks. This is a different approach to measuring ancient temperatures, and its results can be compared with those obtained from studying fossil plants.

Figure 10.9 *Nothofagus*, the southern beech, was one of the common trees in these early and middle Cainozoic forests. They survive today only in moist environments, such as on the slopes of Mt Bulla in Victoria, the fiord country of New Zealand's South Island, and the highlands of New Guinea.

So, the general picture that has been painted for the time between 65 and 20 million years ago is that central Australia was much wetter than now and was covered mainly by rainforest, with a few open areas, and there were large rivers and permanent lakes.

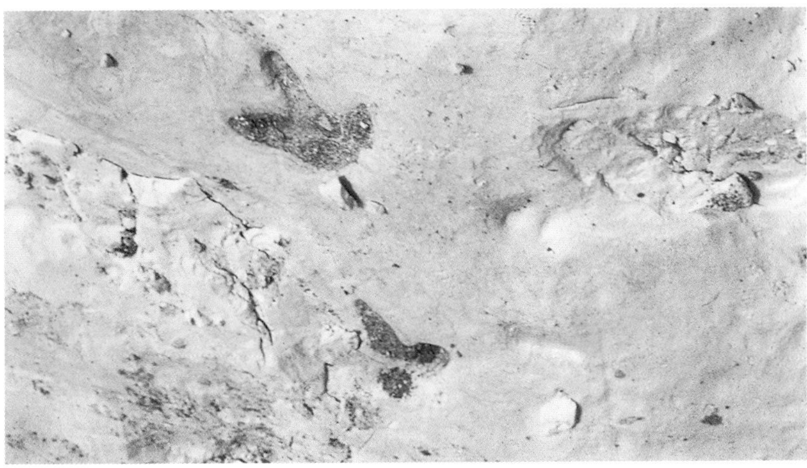

Figure 10.10. Tracks of a late Oligocene bird that was about as high as an emu but decidedly more stout. Tracks were preserved in clays that also contained spores and pollen, which indicate that they belong to the late Oligocene. The tracks were discovered when the clays were being mined near Pioneer, Tasmania, for use in water-filters. (R. Green)

BACKBONED CREATURES

What now of the backboned animals? We haven't a clue what they were doing during the first half of the Cainozoic! Our record of Cainozoic vertebrates essentially begins about 50 million years ago. A single locale near Murgon in Queensland has yielded a handful of tiny teeth of ancient marsupials and a few other weird animals. Another locality at Geilston Bay in Tasmania and another on the north coast of Tasmania near Wynyard, have given us another glimpse of old Cainozoic vertebrates.

Geilston Bay has fossils about 22 to 23 million years old - native cats, possum-like marsupials and diprotodonts, a group that contains a wide variety of today's marsupials such as kangaroos and koalas, but also extinct forms such as *Diprotodon*. Wynyard's only fossil land vertebrate, *Wynyardia*, was a primitive marsupial that looked like a cross between wombats, kangaroos and some other diprotodonts. It was the great, great ..... grandfather or grandmother or a near relative of many of Australia's living diprotodont marsupials. So, we really don't know much about what was happening with the vertebrates on land during the early Tertiary.

Figure 10.11. Geilston Bay, near Hobart in Tasmania, where some of Australia's oldest mammals, Oligocene in age, were dug up last century. The handful of marsupial teeth came from a quarry that is now covered up by a school sports oval. We wonder how many more fossils are still buried there.

The oldest mammalian fossils from central Australia, from the present desert area, are for the most part similar to many of our living marsupials. They come from such places as Lake Palankarinna in South Australia, Riversleigh in Queensland and Bullock Creek in the Northern Territory. The Miocene species, those that lived in central Australia from about 5 to 25 million years old, do tell something about what this country was like then, and the information they provide nicely complements the picture that plant fossils of the same time paint.

In Miocene times, rainforests began to give way to more open vegetation. In the early Miocene, *Nothofagus*, the southern beeches, were still abundant and widespread - and so, we

Figure 10.12. *Wynyardia*, the oldest marsupial with a skeleton preserved, has a skull with no teeth preserved in it. Oligocene in age, it was dug up last century. This fossil was found in marine rocks along the north coast of Tasmania. Evidently its skeleton had been washed out to sea after it died. Its age is known because it is preserved with shells of marine snails and bivalves (clams) that can be quite precisely dated.

Figure 10.13. Wynyard cliffs along the north coast of Tasmania, where *Wynyardia* was found last century. The rocks are late Oligocene in age.

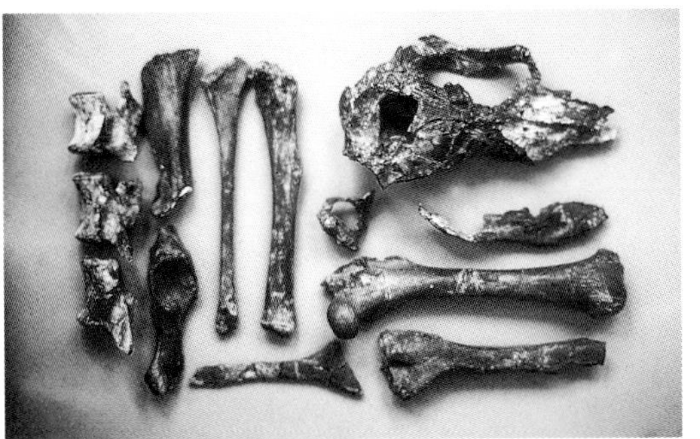

Figure 10.14. Bones of *Wynyardia* — everything but teeth.

Figure 10.15. *Wynyardia*, a mid-Tertiary rather primitive marsupial found at Wynyard in Tasmania in rocks that were laid down in a shallow sea. This carcass had floated out to sea after the animal died, for it most definitely did not live in the ocean. (Painting by F. Knight, courtesy of the Museum of Victoria. From Rich, van Tets and Knight, 1985)

Figure 10.16. Lake Palankarinna, a dry salt pan in northern South Australia, was the first place where Tertiary land-dwelling vertebrates were found in large numbers. R. A. Stirton and his companions from the South Australian Museum and the University of California found fossils at the lake in the early 1950s when they climbed up a sand dune to have their lunch - in the wind and away from the flies. So, perhaps you could say the flies were important in these discoveries!

Figure 10.17. Many Cainozoic vertebrate fossil localities are remote. This sign to Broken Hill is near one of the fossil vertebrate sites in central Australia - and it's a fair way to the nearest milk bar or pub.

Figure 10.18. An international crew at Bullock Creek looking over a field map. (C. Hann)

Figure 10.19. Beating the flies to lunch. Looking for fossils can require some ingenuity! (R. H. Tedford)

Figure 10.20a-e. Excavating fossil-bearing sands and silts in northern South Australia and bagging them in hessian sacks for moving to a bore drain where they are processed. Because the nearest museums are so far away, fossils in sand and clay need to be removed from the surrounding matrix - otherwise you would have to bring back tonnes of rock.

Figure 10.21. Picking matrix back in the laboratory. The small bones and teeth are sorted out from rocks and crystals, sometimes under a microscope. (F. Coffa)

Figure 10.22. Drilling holes in the hard limestones of northern Australia.

(a) Bullock Creek, Northern Territory

(b) Bullock Creek, Northern Territory

(c) Riversleigh, Queensland.

conclude that the rainforests were alive and well at this time. Fossil grass pollen tells us that grasslands grew at least on the mudflats of the rivers that still flowed freely in these central Australian plains. The first wattles (*Acacia*) arrived in Australia from Asia at least by the early Miocene. Australia, after having said its final good-byes to Antarctica at the beginning of the Cainozoic, was now close enough to Asia to receive immigrants from the north. In the middle and late Miocene, gums (*Eucalyptus*) and she-oaks (*Casuarina*) appear for the first time. The countryside was drying out. By the end of the Miocene, *Nothofagus* was gone from the Centre, although it was still wetter there than it is now.

CLUES TO ONCE WETTER CLIMATE

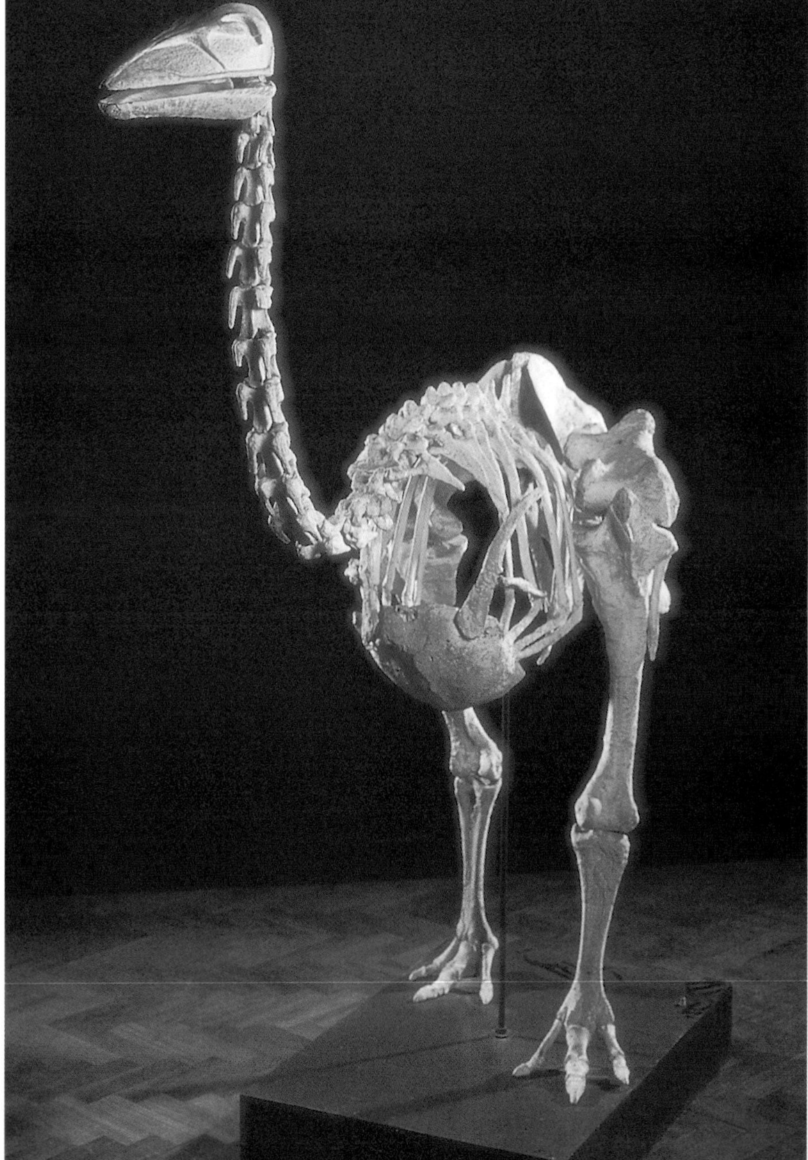

Figure 10.23. *Genyornis*, a giant bird *not* related to the emu, from Lake Callabonna, South Australia. Many species of mihirungs once lived in Australia, but seem not to have adjusted to the increasing openness of the country. The last was gone by the time Captain Cook arrived. (F. Coffa)

Figure 10.24. These bones of the giant mihirung birds in limestone at Bullock Creek, Northern Territory, are Miocene in age.

Figure 10.25. An egg of a giant bird, perhaps a mihirung but more likely that of an Elephant Bird from Madagascar. Shell bits of an even larger egg have been found near Maree in South Australia, an egg big enough to make an omelette for ten people!

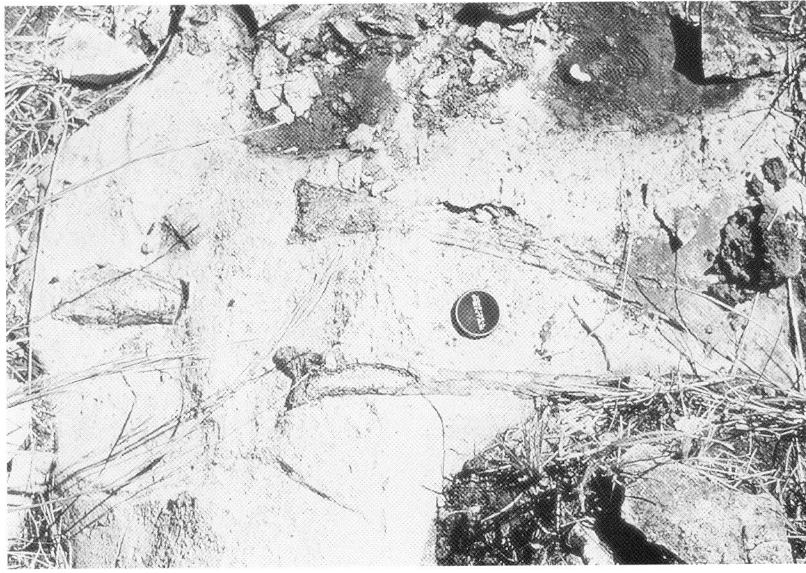

The fossil vertebrate animals of the Miocene also show that the countryside was much wetter then than now. Lungfish, teleost fish and crocodiles were quite common in the lakes and streams, which must have been permanent in order to support this life. Flamingoes and the flamingo-like palaelodid birds were abundant, and they could not have existed unless permanent water was available. Birds that may have been ancestors to both emus and cassowaries haunted the forests of northern South Australia, as did huge forest-dwelling birds, mihirungs, whose nearest modern relatives are probably the fowl-like birds, for example, chickens, pheasants, scrub turkeys, etc. One species of mihirungs did develop long legs that would have been ideal for running in open country (they may have been one of the first vertebrates to moved into the grassy mudflats), but most mihirungs kept to proportions suitable for species that live in forested environments, shorter, squatter legs.

Many of the marsupials living in central Australia in Miocene times spent a lot of time in trees (arboreal animals) - koalas, squirrel-like ektopodontids and some of the possums. Kangaroos of this time had teeth that were adapted to eating soft leaves - they were browsers. Many kangaroos today prefer grasses, and their teeth and jaws have a special design to help them feed on the tough, silicious grass stems, which wear teeth much faster than leaves generally do.

Platypuses swam in the streams and lakes, as did freshwater dolphins. Both animals relied on the lakes and rivers being there all the time. Dolphins don't like sand dunes!

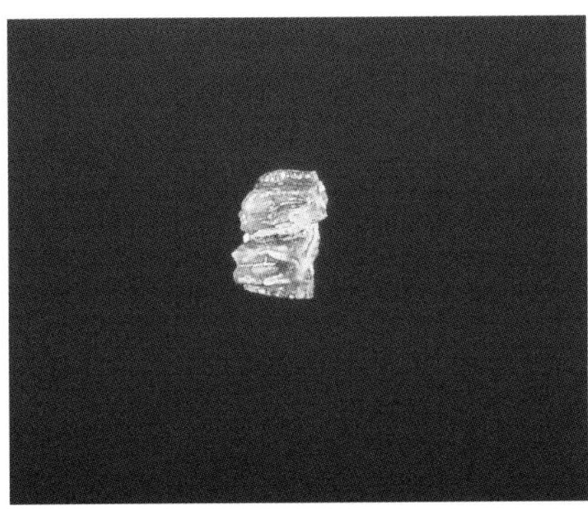

Figure 10.26. The tooth of a platypus. Living platypuses don't have teeth now, just leathery pads in their jaws. Very young ones have weak teeth that have very soft dentine covering them. The owner of this 15-million-year-old tooth from central Australia had proper hard teeth like you and me, with both dentine and enamel on them. The pattern of the teeth was much like the weak teeth of today's platypuses. The black and white squares are 1 centimetre long.

CHANGING CLIMATE AND EXTINCTIONS

Climate changed in the late Cainozic, especially during the late Pliocene and Quaternary. First gums (*Eucalyptus*) became very abundant in central Australia, but then they were crowded out by the blue-bush/salt-bush chenopods, wattles and spinifex that seem to cope better than gums with really dry climates. Open country vertebrate species become dominant in central Australia at this time too. Grazing, grass-eating kangaroos were abundant, and browsing species retreated with the gum forests away from the dry Centre. Dolphins and platypuses disappeared, as many of the rivers no longer flowed all the time. Flamingoes, as well as palaelodid birds, still existed into the Quaternary, but some time during the last 2 million years, even they became extinct. The *permanent* lakes finally dried up. Water filled them only occasionally, but not often enough. Flamingoes had no place to feed or breed on a regular basis. Lungfish and crocodiles disappeared too. The big dry had finally come.

At the same time, temperatures cooled towards the Ice Age we have today, and with this cooling came massive changes in the amount of water that was available from year to year. During the last 700,000 years variable, unpredictable climate has been normal, just the opposite of the even, predictable conditions of the early Cainozoic.

Climatic change in Australia was related to the increase and decrease in the size of the ice sheets at the poles. During the build-up of ice, times were drier, sand dunes were forming. Between 26,000 and 14,000 years ago was the driest period during the whole Cainozoic in Australia. It was at this time that you could walk to Tasmania and New Guinea. This was also when many species became extinct. Many extinctions were undoubtedly because of the dwindling water and food supply caused by aridity; some extinctions may have been brought about by humans, who had been in Australia from at least 40,000 years ago, quite possibly much longer.

If Aboriginal peoples were the cause of some extinctions, which seems likely, they were aiding the forces that had brought about major changes in the environment - those which caused the shift from rainforests to deserts. Groups that did well in rainforests often did not adapt to the expanding open country. Some did, however, and prospered. Mihirungs, flamingoes and diprotodonts became extinct in late Quaternary times, but kangaroos and emus lived on at their expense. So, too, did the seed eaters, the parrots who had evidently been in Australia for a long time and the murid rodents, which had immigrated to the continent only five million years or so before.

Aridity has shaped much of Australia's living fauna. So, too, has our drift into tropical latitudes. Our isolation has also given us a very unusual collection of animals - when compared to those of the rest of the world. There are still leftovers from a different climate that ruled in the early Cainozoic, and those can be found in the rainforests of eastern Australia and New Guinea - places where you can walk back in time!

Figure 10.27. Dolphins like this swam in the flowing streams and lakes of central Australia during Miocene times. (Drawing by F. Knight, courtesy of the Museum of Victoria. From Rich, van Tets and Knight, 1985)

Figure 10.28. The Hamilton locality on the Grange Burn in western Victoria is an unusual site producing fossil vertebrates - its age in years is known. The volcanic rock, a basalt, that covered the river deposits in which the fossils were found has been dated by studying the radioactive elements in it. Under the vertebrate-bearing clays are marine rocks with lots of shells, which means that the vertebrates must be younger than the marine fossils. So we know that the age of this collection of vertebrate bones is about 4.5 million years, plus or minus a few years!

Figure 10.29. Lake Callabonna, South Australia, is a very flat place, but underneath the salt-covered surface lie skeletons of strange marsupials and birds, now extinct, and more than 40,000 years old.

THE BIG DRY

Figure 10.30. Skeleton of a *Diprotodon*, the world's largest marsupial, weathering out on the surface of Lake Callabonna, South Australia

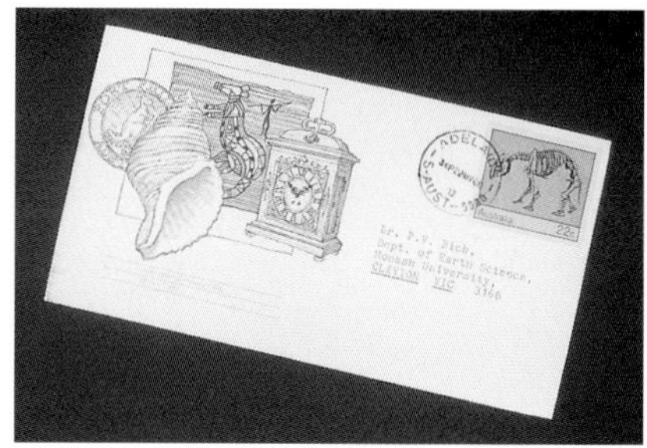

Figure 10.31. *Diprotodon* is one of Australia's historic treasures, as shown on this stamp. (S. Morton)

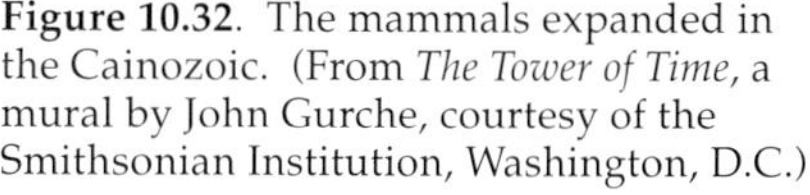

Figure 10.32. The mammals expanded in the Cainozoic. (From *The Tower of Time*, a mural by John Gurche, courtesy of the Smithsonian Institution, Washington, D.C.)

Figure 10.33. A rib of a *Diprotodon* from Cox's Creek, Mullaley, NSW. The hole in this bone may have been made by an Aboriginal tool. *Diprotodon* is now extinct, and just how much humans had to do with killing off this genus is still a big question. (A. Ritchie, R. Wright)

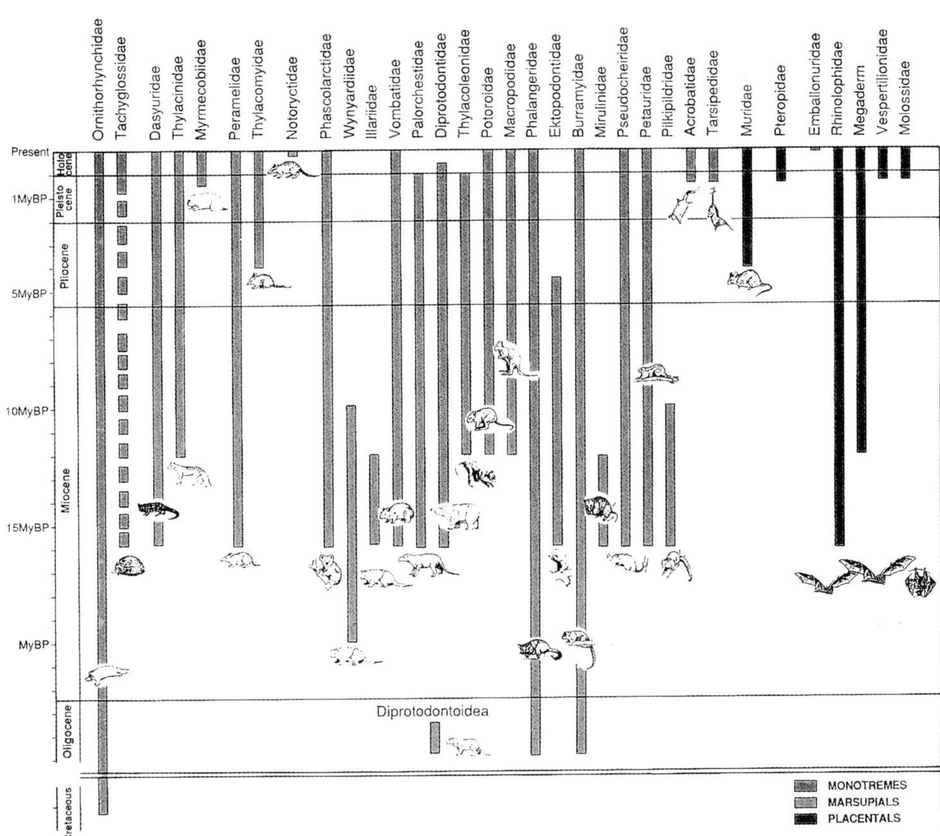

Figure 10.34. Time ranges of the different groups of mammals in Australia. Marsupials are the most important group here, and placentals, like humans, bats, and rats are not at all varied. This is unusual for most of the rest of the world, except South America. (D. Gelt, S. Morton)

Figure 10.35. Tropical butterflies are only a very recent addition to Australia, because, over the last 300 million years, this country has been in tropical latitudes for only a few million years. The ancestors of these butterflies must have lived in the north, in South-East Asia and Indonesia. (K. Walker)

THE LAST 65 MILLION YEARS

66 to 1.64 million years ago

During the Tertiary, Australia began to look like it does today. In Cretaceous times, Australia began splitting apart from Antarctica, and the Southern Ocean was born. The first ocean waters began to flow between the two continents about 50 to 55 million years ago. Climates cooled all over the world, and big ice sheets began to grow at the poles. Australia was a cool, well-watered land covered with rainforests, but as it drifted northwards, it dried out and *Eucalyptus*, wattles, and open grasslands became the rage. On this isolated island continent birds and marsupials developed that were very different from those anywhere else in the world - emus, cassowaries, kangaroos and koalas. The first placental mammals, bat and rats, arrived in Australia during the last 50 million years, rats only in the last four million.

Figure 10.36. The drying out of Australia in the late Cainozoic brought fire, a familiar part of our modern environment and a tool used by humans to open up the country for hunting animals or to provide space for agriculture.

The Environment

During the early Tertiary, Australia broke away from Antarctica and began drifting northwards. As the earth cooled for a time, Australia warmed as it moved from the polar region towards the tropics. Warm waters of the Indian Ocean flowed eastwards, then along southern Australia, at first forming a deep gulf like the Red Sea and finally an open, ever widening ocean about 30 to 40 million years ago. Ocean currents encouraged the chilling of Antarctica, as warm waters from the north were blocked from moving far south.

The Tasman Sea had opened in the late Cretaceous, beginning 80 to 90 million years ago. Later the Coral Sea opened. Volcanoes erupted down the east coast of Australia, somehow related to the continent's northward drift. The eastern highlands, Australia's highest mountains, also began to rise in the late Cretaceous and continued to do so through the Cainozoic.

Australia began to collide with lands to the north, which caused intense mountain building about 12 to 15 million years ago. Modern New Guinea began to emerge from the ocean bottom.

Through most of the Cainozoic, Australia was an island continent, which like a "Noah's Ark" moved from Antarctica towards Asia with its *menagerie* of animals and plants. They have had a history very separate from animals in the rest of the world.

The Plants

Early Tertiary plants lived in a cool, well-watered climate, forming rainforests similar to those in Antarctica and New Zealand of the same age and somewhat like those still growing on New Zealand's South Island today. Coal swamps covered parts of southern Australia and stayed there as long as humid, cool conditions prevailed. These coals are a major power source for Melbourne and other parts of Australia's southeast. *Nothofagus*, the southern beeches, and a variety of flowering plants, podocarps and araucarian conifers, as well as ferns were abundant in these coal-forming forests. By Miocene times, the abundant in these coal-forming forests. By Miocene times, the drying out of central Australia had caused big changes in the plants. Gum trees (*Eucalyptus*) developed and wattles (*Acacia*) invaded from the north as Australia came closer to South East Asia. Fossils of fruits in siliceous rocks of central Australia show that eucalypts existed by the beginning of the Miocene, 20 to 25 million years ago.

The Backboned Animals

Vertebrates are abundant fossils in Tertiary sands, but the record of land dwelling forms is only well known from the last 25 million years. Freshwater fish are known in slightly older deposits. The Eocene - Oligocene Redbank Plains sediments of Queensland and the Rundle Oil shales have fossil fish somewhat like those in the Eocene Green River Formation of North America - advanced bony fish, the teleosts. The living lungfish, *Neoceratodus*, also lived during these times. The Miocene lake deposits in the Warrumbungle Mountains of central New South Wales contain fossil fish closely related to the Murray Cod, *Maccullochella*, as well as a nearly complete specimen of a small bird, the owlet-nightjar *Quipollornis*. Marine fish are known from much of the Tertiary, including a great number of sharks and ray-finned fish. These are similar to living forms in Australian coastal waters, but there were also a few subtropical forms that are now extinct.

Figure 10.37. The world and Australia in the early Tertiary. The continent was much wetter, and rivers flowed in many parts of central Australia where there is desert today. Much of the eastern coastal part was bristling with erupting volcanoes. For part of this time there was a dry-land connection with lands to the south, but about 50 million years ago Australia and Antarctica finally parted company. The Southern Ocean broke through to the growing Tasman Sea and made Australia an island continent.

When land-living backboned animals appeared in the Teritary fossil record of Australia, most looked a lot like groups that are living here now. The wynyardiids are one of the few

Figure 10.38. Even in the middle Tertiary, about 15 million years ago, Australia was still wetter than it is now. Our deserts are very recent, most only beginning to appear five to seven million years ago - and real dryness developed only a few hundred thousand years ago.

ancient groups that show a linkage between living families, such as wombats and kangaroos, and are apparently the remnants of much more primitive marsupials that were abundant in the early Cainozoic. These groups seem to have developed from a common ancestor. Some groups such as the squirrel-like, tree-dwelling Ektopodontidae are so unusual that their nearest relatives are not really known. But, for the most part, modern families of marsupials were around even in mid-Tertiary times - kangaroos, koalas, possums, native cats. Bird families, are also quite modern by 15 million years ago, but some groups are unusual. These have left no survivors in Australia, such as the flamingoes (*Phoeniconotius*) and the flamingo-like palaelodids, which lived in the permanent lakes of central Australia alongside freshwater dolphins and toothed platypuses. Australian Miocene palaelodids were related to French species of similar age, as were bats, the oldest placental mammals known in Australia. Late in Tertiary times, murid rodents invaded our continent from the north, about 4 million years ago.

Figure 10.39a-c. The brown coal mine at Morwell near Melbourne. Melbourne's electricity is provided by burning this coal which was produced by the decay and burying of early Cainozoic swamp forests. In these coals are fireholes, created when the coal caught fire in Pleistocene times and holes were burned in the coal seams. Water filled these holes to form lakes, animals fell into them and drowned, and their skeletons were preserved in the lake muds. Many have been dug up, and some are on display at the Morwell power station museum. (C. Tassell)

Figure 10.40. Some fossil lungfish from the Cainozoic of Australia. *Ceratodus wollastoni* is hiding in the weeds; *Neoceratodus gregoryi* is in front, and two *Neoceratodus eyrensis* are swimming above. Many of today's dried-out lakes and rivers in central Australia were home to these plant-eating fish 10 to 20 million years ago. (Painting by F. Knight, courtesy of the Museum of Victoria. From Rich, van Tets and Knight, 1985)

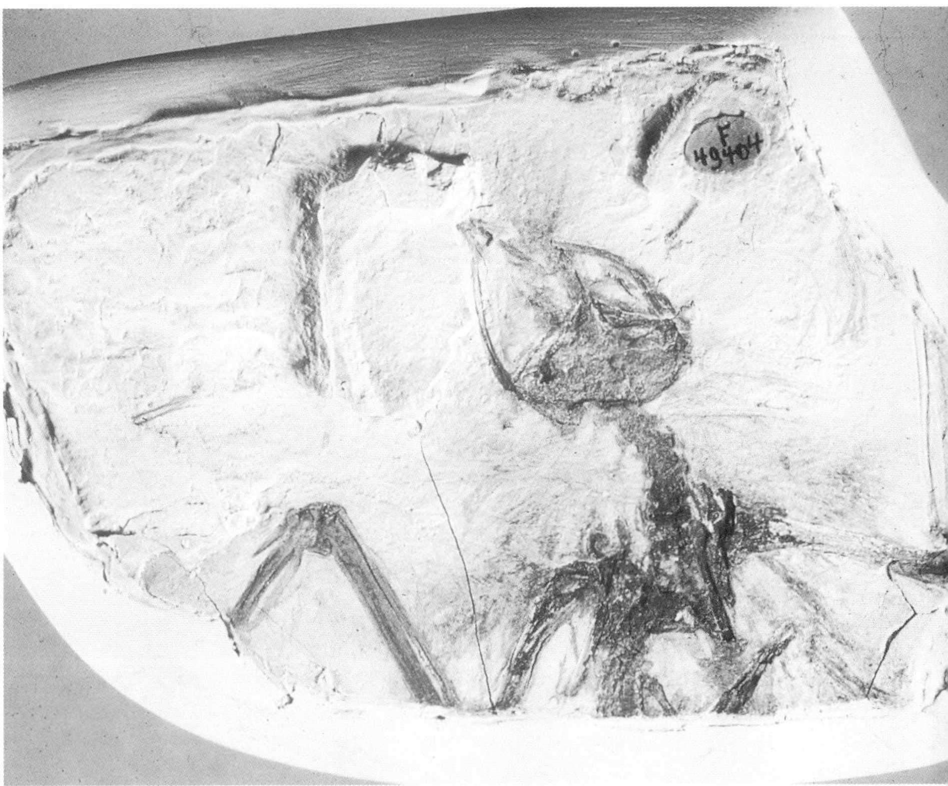

Figure 10.41. The skeleton of *Quipollornis*, a little bird about the size of a nightjar or whippoorwill. It lived in Australia about 15 million years ago where the Warrumbungle Mountains, NSW, are today. It is unusual because it is so complete - its whole skeleton and even its feathers are preserved. (F. Coffa)

Figure 10.42. An impression of *Quipollornis* in flight 15 million years ago. (Drawing by F. Knight, courtesy of the Museum of Victoria. From Rich, van Tets and Knight, 1985)

Although Miocene terrestrial vertebrates in Australia were forest dwellers, a few forms, even in mid-Tertiary times, preferred the open country. By the end of the Tertiary, however, most central Australian vertebrates were open country dwellers, which preferred grazing to browsing. Many had long legs and small toes on the outside of the foot, such as those of emus and kangaroos. The grasslands and their dwellers were thriving under these increasingly dry conditions.

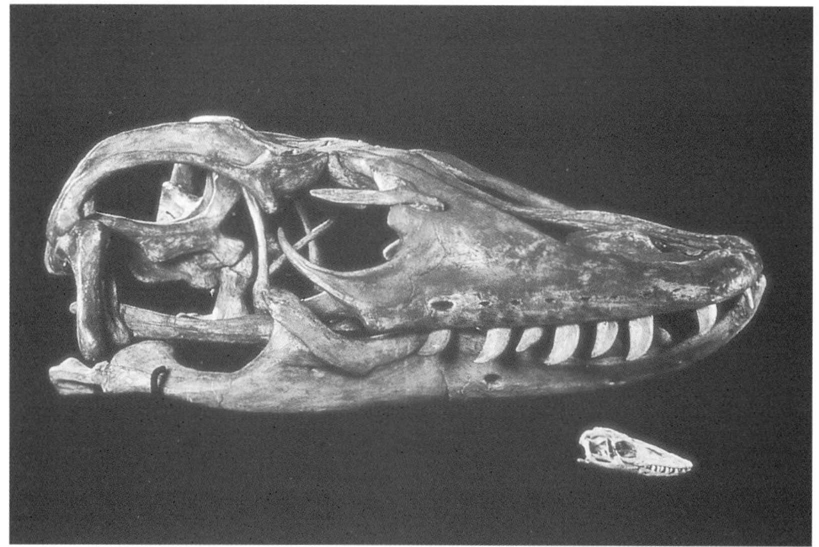

Figure 10.43. Skull of *Megalania* compared to a modern goanna skull that is about 6 centimetres long. *Megalania* grew to about 7 metres long and was the largest carnivorous backboned animal in Cainozoic Australia.

Figure 10.44. *Ektopodon*, a strange-toothed marsupial that lived in the cool rainforests of central Australia during the Miocene. It may have been very similar to a squirrel in habits - a marsupial squirrel, however! (Drawing by F. Knight, courtesy of the Museum of Victoria. From Rich, van Tets and Knight, 1985)

Figure 10.45. Flamingoes and the flamingo-like palaelodids (white birds) were once living on central Australian lakes by the thousands. This is an impression of a scene about 15 million years ago at Lake Palankarinna in northern South Australia. (Drawing by F. Knight, courtesy of the Museum of Victoria. From Rich, van Tets and Knight, 1985)

THE QUATERNARY WORLD
The last 1.64 million years

Several ice ages happened during the last 1.64 million years, and this resulted in several periods of extreme dryness and very low sea levels. The driest time in Australia was about 20,000 years ago. During this time plants were (and still are) mainly flowering plants, and grasslands covered lots of the continent. Species in many groups of backboned animals became very large and by the end of the period many had become extinct. Aborigines reached Australia at least 40,000 years ago and overlapped in time with many of the now extinct giant forms, the megafauna.

Figure 10.46 a-b. Caves are a good place to preserve fossils and give us a good idea of animals that existed in the Pleistocene to Recent in Australia. The bones got into the caves in many ways! (J. Hope)

AUSTRALIA IN AN ICE AGE

The Environment

As the icecap in Antarctica grew, wind circulation patterns began to look like those of today. Anticlockwise air circulation developed at high latitudes, and moved northwards with increasing cold, eventually catching up with Australia and causing really extreme dryness. Northern Australia came under the influence of a subtropical climate as Australia moved north. Polar ice caps grew and shrank and in doing so controlled climate, causing sea levels to rise and fall several times. Under these colder, drier, windier conditions the great long dunes of central Australia formed. Periods of glaciation produced the driest conditions and the lowest sea levels, sometimes as much as 200 metres below what it is now. Tasmania and New Guinea were connected to mainland Australia by land during these times, and the Great Barrier Reef at times became dry land.

The Plants

Climatic changes, more radical and rapid than in the Tertiary, controlled which plants grew and where. Closed forest gave way to woodlands, then to grasslands and, in some cases, deserts. Vegetation adapted to fire. The growth and shrinkage of the ice sheets meant only a change in vegetation patterns in Australia, not so much extinction of species, as no big ice sheets ever formed on this continent during the Quaternary. It was this change that led to our modern drought- adapted plants - eucalypts, acacias, and spinifex. Those plants left over from wetter, cooler times, such as the *Nothofagus* and ferns, lived on only in isolated spots of rainforest along the east and southeast coasts.

Figure 10.46c. Brent Hall putting the final touches on the reconstructed skeleton of *Megalania*. This gives some idea of just how big this giant goanna was. It could have made a good meal out of Brent. (F. Coffa)

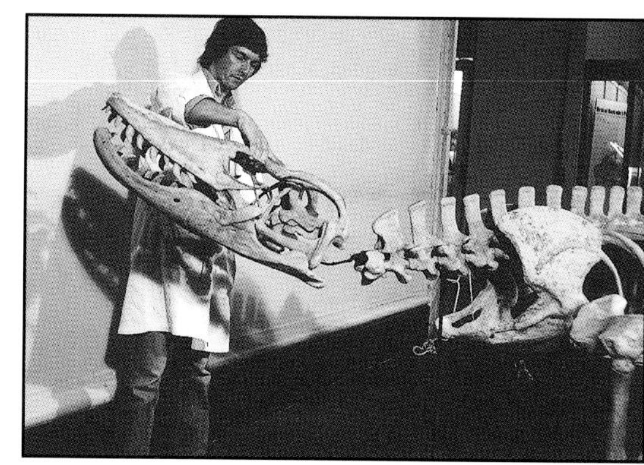

The Backboned Animals

Quaternary vertebrates were a mixture of some species that survived to the present and others that were soon to become extinct. Some forms such as kangaroos (*Procoptodon*, *Sthenurus*,

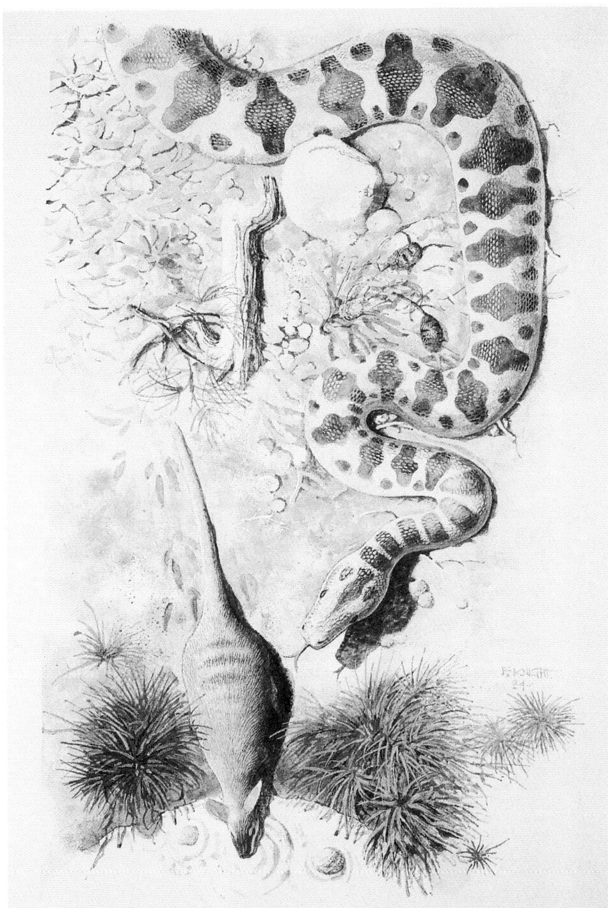

Figure 10.47 a-b. Giant reptiles are known from the Pleistocene of Australia.

a. *Wonambi*, a giant snake, is known from Australia's southeast. (Drawing by F. Knight, courtesy of the Museum of Victoria. From Rich, van Tets and Knight, 1985)

b. The big lizard, *Megalania*, which grew to 6 or 7 metres in length, is known from many places in central and eastern Australia. (Drawing by F. Knight, courtesy of the Museum of Victoria. From Rich, van Tets and Knight, 1985)

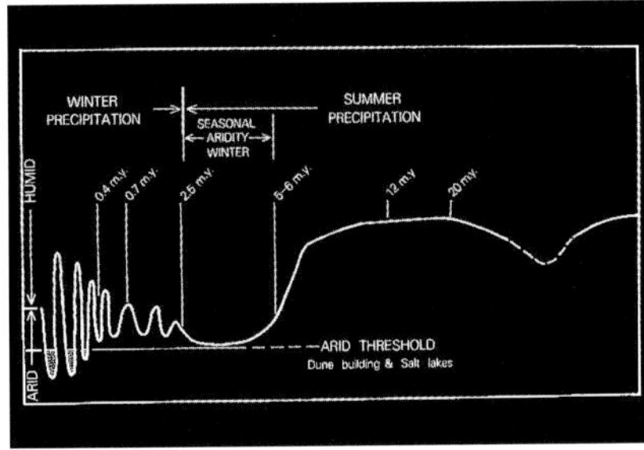

Figure 10.48. The climatic cycles that led to great dryness in Australia began only about half a million years ago. (Modified from J. Bowler)

Macropus), diprotodontid marsupials (*Diprotodon*), emus and varanid lizards (*Megalania*) reached gigantic sizes during the Quaternary. Perhaps greater size gave them a slightly better chance to survive in times of drought, an ever present danger in this time period.

Central Australia still must have supported some permanent fresh-water lakes, because flamingoes lived there for a while in the Quaternary; finally, they died out, leaving the fickle, unpredictable climate to the emu, grazing kangaroos and growing mobs of rodents and parrots.

An unusual aspect of Quaternary (and for that matter, Tertiary) vertebrate faunas is the lack of large carnivores. The puma-sized marsupial lion (*Thylacoleo*) and Thylacine Wolf (*Thylacinus*) were the largest mammalian predators. *Megalania*, a 5 to 7 metre long varanid lizard was the largest "carnivore". Most likely the only truly open country running carnivore was *Propleopus*, a carnivorous kangaroo!

Humans arrived in Australia relatively recently, sometime between 40,000 and 100,000 years ago. Their effects - firing and clearing the land, killing local species and bringing in new ones such as the dingo and rabbits, has had, and continues to have, a big effect on our unique fauna and flora. The overall result of this intervention will only be known in centuries to come.

CHAPTER 11:
TOMORROW

Australia's Face and Place in the Future. Kangaroos and Pandas, Eucalypts and Birch - Which Will Win?

The next 50 million years

Douglas stepped into his 1998 Holden, set the dial for 50 million plus and strapped himself firmly in the driver's seat. He looked around at the rolling coastal hills that suddenly dropped away 100 metres below to the Southern Ocean. Out there was nothing but blue sea to the horizon. The crows were beginning to call as darkness settled in. The cool of the evening breeze washed across his face. He took a last survey around him, and drew a long breath of the sweet, salty air. A red button marked 'Escape Velocity' glowed dimly on the control panel. His index finger touched it, hesitating, then pushed it firmly down. Douglas and his car promptly disappeared!

A TREK TO THE EQUATOR

Some people, especially science fiction lovers, dream of heading off into the future in a time machine. Some even prefer going backwards in time. Whichever way you go, it would be fantastic to step out of our own time and go for a look at some other age - that is, of course, providing we had a choice of coming back, and living through the trip! If we could time travel, it would certainly be easier to work out the colour of *Tyrannosaurus* or the reasons why so many animals and plants became extinct about 65 million years ago, or what all the pollution in our atmosphere would lead to. But life is not so easy. So far, despite our dreams and determined efforts, no one has *yet* built a time machine.

So, if we don't have a time machine, how can we go about looking into the future. Try a crystal ball? Not quite! There are other ways. We can start by carefully observing how things work today, in detail. Many of the same things that happen today have happened in the past. The geologic record of rocks and fossils tells us that. There are, of course, some things that have happened in that past that aren't still happening, at least right now. For example, we know the atmosphere two billion years ago had a lot less oxygen in it than it does now. We know this because scientists have studied the kinds of minerals that were being deposited in the rivers and lakes of that far distant time. But as far as we know the physical laws that govern the earth and the universe applied in the past just as they do now and will in the future.

Using all we know about past and present, some *educated* predictions about Australia's future can be made. Fifty million years in the future we would most likely feel at home with the plants and geography. There will be some changes, though. If Australia continues to move northwards at its present rate, it will have smashed into Southeast Asia, and dry land will connect China and New Guinea. That dry land will probably be a major mountain range similar to today's Himalayas, created as the two continents, Asia and Australia, collide. The Red Sea will be a lot wider and a part of east Africa will have split off from its mother continent and be drifting eastward, forming a big island. South America and North America will no longer be connected by a bridge of land, so there will be no need at all for the Panama Canal. A natural waterway will slice the two continents apart. To the north, a bridge of mountainous land will connect North America and Asia, and maybe a big mountain range will be the 'glue' between North Africa and Europe. Only South America and Antarctica will be isolated, island continents.

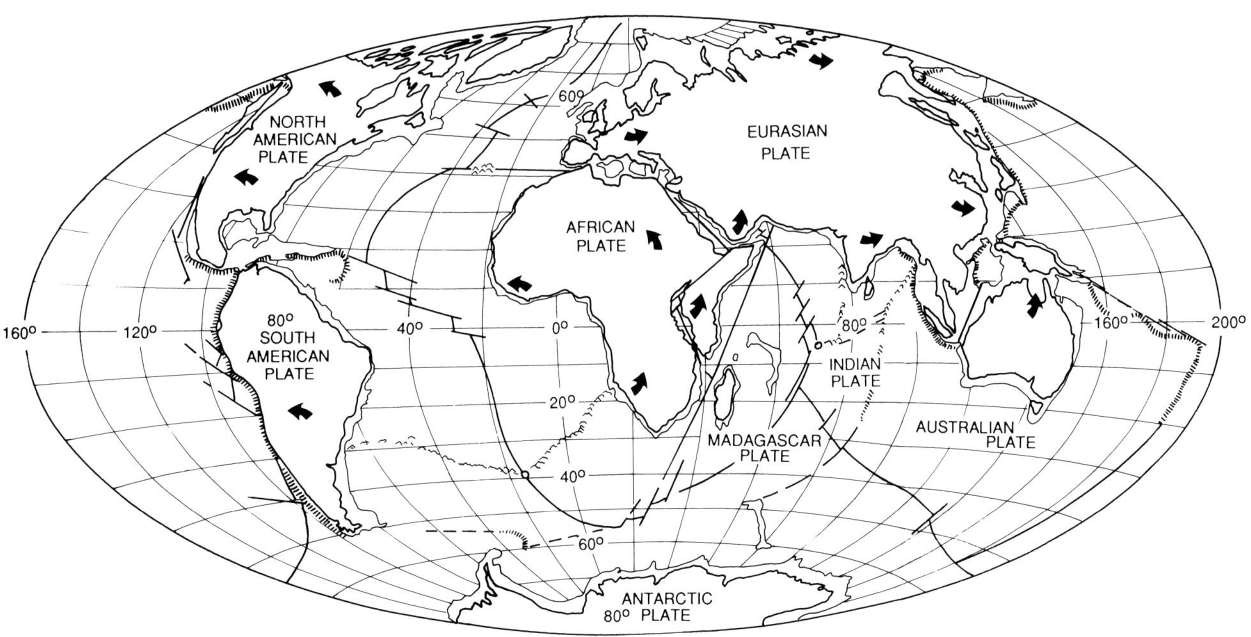

Figure 11.1. The world as it may look in 50 million years time. (Modified by D. Gelt from map by Dietz and Holden, (1970), pp. 30-41).

And the climate? That will depend on many things, not the least being human beings. Will the pollution that we are pouring into the air cause a runaway greenhouse effect, raising the temperature up to levels like those observed on the surface of Venus, too hot for life as we know it? Or will we lose the ozone layer, depriving ourselves and all life on the surface of the earth of protection from the ultraviolet rays of the sun? Or will humans do something that drives away the atmosphere, to leave our planet as cold as Mars? Will humans face reality or become extinct and leave the protective atmosphere of the earth

WHEN AUSTRALIA AND ASIA MEET

about the same as it is now? That we cannot foresee, just yet. Our models based on present and past offer many possibilities but no *single* solution!

Fifty million years in the future - if our earthly climates remain within the range we have known for the past 500 million years - Australia should have a tropical climate, for it will lie astride the equator. Rainforests, warm and humid in nature, will be widespread across the country, and a great variety of climbing marsupials will live at all levels in this lush vegetation. Tree kangaroos will probably be quite varied and may be pursued by leopard-like, largish native cats. Although most tree dwelling kangaroos will be plant eaters, some may indeed have become blood-letting killers, just as some grassland forms in the late Cainozoic did. Perhaps a new carnivore will develop from the tree-loving possums - clever and fast moving in the trees. On the ground, panda-like marsupials may feed heartily on the bamboo-like grasses.

The mammals of this continent will still be mainly marsupials - for, despite the fact that Australia will have crashed into Asia, a mighty mountain range, perhaps higher than the magnificent Himalayas of today, will serve as a barrier to all but the most hearty of the northern invaders. Before reaching the rich tropical land of Australia, these 9000 metre mountains with their frigid peaks will have to be dealt with.

With all kinds of possibilities, your imagination could run wild reconstructing the future forest animals and plants of Australia's tropics. Parrots might well become more varied than now, moving into niches or lifestyles more typical of rodents elsewhere in the world - many may have become burrowers, some may even be carnivores. This may be the success before the fall, however, for reproductive rates of rodents tend to be higher than those of parrots. In the end they may leave the rodents superior, once the connections with Asia are more passable, once the mountains have eroded down, but then maybe not.

The Great Barrier Reef will span the whole east coast of Australia in this time 50 million years in the future, and perhaps similar reefs will develop at places along the west coast.

The stage would be set for a future battle, when the flow of animals and plants could take place between the great continent 'Norafeuasia' (an amalgam of North America, Africa, Europe and Asia) and the isolated peninsula, Australia. If the past is any guide to the future, unique Australian animals, such as the marsupial koalas and kangaroos, the native cats and mice may not do so well against the invading placentals. When North America and South America exchanged animals in the late Cainozoic, the devastation of South American marsupials and primitive placentals was terrible. We also know that some parrots live in South-East Asia and Africa today but have never been particularly varied there. What has caused this rarity? Are they competing, rather poorly, with some other groups in their environment, perhaps rodents? And if so, will the varied parrots, so much a part of Australian life, meet with a similar

AND THEN WHAT ?

fate in these future times when northern faunas meet those of Australia?

There are so many variables, so many possibilities that our armchair time machine is not good enough to help us choose which is the most likely. We're not sure that humans will even still be around - but to increase our chances, one bit of positive action is worth considering. We should make an attempt not to tamper too severely with our atmosphere. Once the atmosphere is lost, life on a world scale would be reduced to the basics, and humans are not likely to be a part of that.

Together with maintaining and nurturing our atmosphere, we need to continue our quest to understand the past and present ecosystems and define factors that have caused them to change, to collapse or to prosper. This is our best bet for making accurate predictions about our future. This quest is the best crystal ball we have, the best time machine we can find, and perhaps our only chance at eternity.

SOME REFERENCES FOR FURTHER READING

Archer, M., Hand, S.J. and Godthelp, H., 1991. *Riversleigh. The Story of Animals in Ancient Rainforests of Inland Australia.* Reed Books International, Sydney.

Attenborough, D. 1979. *Life on Earth.* Collins, London.

Beerbower, J. R. 1968. *Search for the Past: An Introduction to Paleontology.* Prentice Hall. Englewood Cliffs, New Jersey.

Clark, I. F. and Cook, B. J. (eds.) 1986. *Geological Science. Perspectives of the Earth.* Australian Academy of Science, Canberra.

Colbert, E. H. and Morales, M. 1991. *Evolution of the Vertebrates.* 4th Ed. Wiley–Liss Inc., New York.

Desmond, A. J. 1977. *The Hot-Blooded Dinosaurs.* Futura Publ. Ltd., London.

Eicher, D. L. and McAlester, A. L. 1980. *History of the Earth.* Prentice Hall, Englewood Cliffs, New Jersey.

Flannery, T. F., 1994. *The Future Eaters.* Reed Books International, Sydney.

Hamilton, R. 1978. *Fossils and Fossil Collecting.* Hamlyn, New York.

Hand, S. and Archer, M. 1987. *The Antipodean Ark.* Angus and Robertson, Sydney.

LaPorte, L. F. 1968. *Ancient Environments.* Prentice Hall, Englewood Cliffs, New Jersey

Ley, W. 1968. *Dawn of Zoology.* Prentice Hall, Englewood Cliffs, New Jersey.

McAlester, A. L. 1977. *The History of Life.* Prentice Hall, Englewood Cliffs, New Jersey.

Murray, P. 1984. *Australia's Prehistoric Animals.* Methuen, North Ryde.

Norman, D. 1985. *The Illustrated Encyclopedea of Dinosaurs.* Salamander Books, London.

Quirk, S. and Archer, M. (eds.) 1983. *Prehistoric Animals of Australia.* Australian Museum, Sydney.

Rich, L. S. and Rich, P. V. 1987. *Australian Armoured Fish and Their World (The Palaeozoic).* Golden Project Book, Golden Press, Sydney.

Rich, P. V. 1991. *The Dinosaur Hunter's Kit* and *The Dinosaur Hunter's Handbook.* Five Mile Press, Melbourne.

Rich, P. V. 1987. *Australian Dinosaurs and Their World (The Mesozoic).* Golden Project Book, Golden Press, Sydney.

Rich, P. V. and Rich, L. S. 1987. *Australia's Prehistoric Birds and Carnivorous Kangaroos and Their World (The Cainozoic).* Golden Project Book, Golden Press, Sydney.

Rich, P. V., Rich, T. H., and Fenton, M. A. 1989. *The Fossil Book.* Doubleday, New York.

Rich, P. V., van Tets, G. F., and Knight, F. 1985. *Kadimakara: Extinct Vertebrates of Australia.* Pioneer Design Studio, Lilydale.

Rixon, A. E. 1976. *Fossil Animal Remains. Their Preparation and Conservation.* Athlone Press, Univ. London.

Rudwick, M. J. S. 1985. *The Meaning of Fossils. Episodes in the History of Palaeontology.* 2nd Ed., University of Chicago Press, Chicago and London.

Stanley, S. M. 1984. Mass Extinction in the Ocean. *Scientific American* volume 250, no. 6, pp. 46-54.

Stone, D., Dunnett, G. and Stone, D., 1991. *Explore the Great Ocean Road. Along Australia's Southern Touring Route.* See Australia Guides, Burwood.

Tarling, D. H. and Tarling, M. P. 1975. *Continental Drift.* Pelican, Aylesbury.

Vickers-Rich, P. 1992. *The Dinosaur Bag, An Educational Activity Kit (Ages 5-12).* Monash University Earth Sciences Department, Melbourne.

Vickers-Rich, P., Monaghan, J. M., Baird, R. F. and Rich, T. H. 1991. *Vertebrate Palaeontology of Australasia.* Pioneer Design Studio in cooperation with the Monash University Publications Unit, Melbourne.

Vickers-Rich, P. and Rich, T. H., 1993. *Wildlife of Gondwana,* Reed Books International, Sydney.

Vickers-Rich, P. and Rich, T. H., 1993. Australia's Polar Dinosaurs. *Scientific American* volume 269, no. 1, pp. 50–55.

White, M. E. 1984. *Australia's Prehistoric Plants and Their Environment.* Methuen, North Ryde.

White, M. E. 1986. *The Greening of Gondwana.* Reed Books International, Sydney.